New Security Challenges

General Editor: **Stuart Croft**, Professor of International Security in the Department of Politics and International Studies at the University of Warwick, UK, and Director of the ESRC's New Security Challenges Programme.

The last decade demonstrated that threats to security vary greatly in their causes and manifestations, and that they invite interest and demand responses from the social sciences, civil society and a very broad policy community. In the past, the avoidance of war was the primary objective, but with the end of the Cold War the retention of military defence as the centrepiece of international security agenda became untenable. There has been, therefore, a significant shift in emphasis away from traditional approaches to security to a new agenda that talks of the softer side of security, in terms of human security, economic security and environmental security. The topical *New Security Challenges* series reflects this pressing political and research agenda.

Titles include:

Abdul Haqq Baker
EXTREMISTS IN OUR MIDST
Confronting Terror

Robin Cameron
SUBJECTS OF SECURITY
Domestic Effects of Foreign Policy in the War on Terror

Jon Coaffee, David Murakami Wood and Peter Rogers
THE EVERYDAY RESILIENCE OF THE CITY
How Cities Respond to Terrorism and Disaster

Sharyl Cross, Savo Kentera, R. Craig Nation and Radovan Vukadinovic (*editors*)
SHAPING SOUTH EAST EUROPE'S SECURITY COMMUNITY FOR THE TWENTY-FIRST CENTURY
Trust, Partnership, Integration

Tom Dyson and Theodore Konstadinides
EUROPEAN DEFENCE COOPERATION IN EU LAW AND IR THEORY

Tom Dyson
NEOCLASSICAL REALISM AND DEFENCE REFORM IN POST-COLD WAR EUROPE

Håkan Edström, Janne Haaland Matlary and Magnus Petersson (*editors*)
NATO: THE POWER OF PARTNERSHIPS

Håkan Edström and Dennis Gyllensporre
POLITICAL ASPIRATIONS AND PERILS OF SECURITY
Unpacking the Military Strategy of the United Nations

Hakan Edström and Dennis Gyllensporre (*editors*)
PURSUING STRATEGY
NATO Operations from the Gulf War to Gaddafi

Christopher Farrington (*editor*)
GLOBAL CHANGE, CIVIL SOCIETY AND THE NORTHERN IRELAND PEACE PROCESS
Implementing the Political Settlement

Adrian Gallagher
GENOCIDE AND ITS THREAT TO CONTEMPORARY INTERNATIONAL ORDER

Kevin Gillan, Jenny Pickerill and Frank Webster
ANTI-WAR ACTIVISM
New Media and Protest in the Information Age

James Gow and Ivan Zverzhanovski
SECURITY, DEMOCRACY AND WAR CRIMES
Security Sector Transformation in Serbia

Toni Haastrup
CHARTING TRANSFORMATION THROUGH SECURITY
Contemporary EU-Africa Relations

Ellen Hallams, Luca Ratti and Ben Zyla (*editors*)
NATO BEYOND 9/11
The Transformation of the Atlantic Alliance

Andrew Hill
RE-IMAGINING THE WAR ON TERROR
Seeing, Waiting, Travelling

Christopher Hobbs, Matthew Moran and Daniel Salisbury (*editors*)
OPEN SOURCE INTELLIGENCE IN THE TWENTY-FIRST CENTURY
New Approaches and Opportunities

Andrew Hoskins and Ben O'Loughlin
TELEVISION AND TERROR
Conflicting Times and the Crisis of News Discourse

Paul Jackson and Peter Albrecht
RECONSTRUCTING SECURITY AFTER CONFLICT
Security Sector Reform in Sierra Leone

Bryan Mabee
THE GLOBALIZATION OF SECURITY
State Power, Security Provision and Legitimacy

Janne Haaland Matlary
EUROPEAN UNION SECURITY DYNAMICS
In the New National Interest

Kevork Oskanian
FEAR, WEAKNESS AND POWER IN THE POST-SOVIET SOUTH CAUCASUS
A Theoretical and Empirical Analysis

Michael Pugh, Neil Cooper and Mandy Turner (*editors*)
WHOSE PEACE? CRITICAL PERSPECTIVES ON THE POLITICAL ECONOMY OF PEACEBUILDING

Brian Rappert and Chandré Gould (*editors*)
BIOSECURITY
Origins, Transformations and Practices

Nathan Roger
IMAGE WARFARE IN THE WAR ON TERROR

Aglaya Snetkov and Stephen Aris
THE REGIONAL DIMENSIONS TO SECURITY
Other Sides of Afghanistan

Ali Tekin and Paul Andrew Williams
GEO-POLITICS OF THE EURO-ASIA ENERGY NEXUS
The European Union, Russia and Turkey

Lisa Watanabe
SECURING EUROPE

Mark Webber, James Sperling and Martin A. Smith
NATO's POST-COLD WAR TRAJECTORY
Decline or Regeneration

New Security Challenges Series
Series Standing Order ISBN 978–0–230–00216–6 (hardback) and ISBN 978–0–230–00217–3 (paperback)
(*outside North America only*)

You can receive future titles in this series as they are published by placing a standing order. Please contact your bookseller or, in case of difficulty, write to us at the address below with your name and address, the title of the series and the ISBNs quoted above.

Customer Services Department, Macmillan Distribution Ltd, Houndmills, Basingstoke, Hampshire RG21 6XS, England

Open Source Intelligence in the Twenty-First Century

New Approaches and Opportunities

Edited by

Christopher Hobbs, Matthew Moran and Daniel Salisbury
Centre for Science and Security Studies, King's College London, UK

Selection, introduction, conclusion and editorial matter © Christopher Hobbs, Matthew Moran and Daniel Salisbury 2014
Individual chapters © Respective authors 2014

All rights reserved. No reproduction, copy or transmission of this publication may be made without written permission.

No portion of this publication may be reproduced, copied or transmitted save with written permission or in accordance with the provisions of the Copyright, Designs and Patents Act 1988, or under the terms of any licence permitting limited copying issued by the Copyright Licensing Agency, Saffron House, 6–10 Kirby Street, London EC1N 8TS.

Any person who does any unauthorized act in relation to this publication may be liable to criminal prosecution and civil claims for damages.

The authors have asserted their rights to be identified as the authors of this work in accordance with the Copyright, Designs and Patents Act 1988.

First published 2014 by
PALGRAVE MACMILLAN

Palgrave Macmillan in the UK is an imprint of Macmillan Publishers Limited, registered in England, company number 785998, of Houndmills, Basingstoke, Hampshire RG21 6XS.

Palgrave Macmillan in the US is a division of St Martin's Press LLC,
175 Fifth Avenue, New York, NY 10010.

Palgrave Macmillan is the global academic imprint of the above companies and has companies and representatives throughout the world.

Palgrave® and Macmillan® are registered trademarks in the United States, the United Kingdom, Europe and other countries.

ISBN 978–1–137–35331–3

This book is printed on paper suitable for recycling and made from fully managed and sustained forest sources. Logging, pulping and manufacturing processes are expected to conform to the environmental regulations of the country of origin.

A catalogue record for this book is available from the British Library.

A catalog record for this book is available from the Library of Congress.

Transferred to Digital Printing in 2014

D 10 9 8 7 6

Contents

List of Tables and Figures	vii
Acknowledgements	viii
Notes on Contributors	ix
List of Abbreviations	xii

Introduction   1
Christopher Hobbs, Matthew Moran and Daniel Salisbury

Part I   Open Source Intelligence: New Methods and Approaches

1. Exploring the Role and Value of Open Source Intelligence   9
  Stevyn D. Gibson

2. Towards the Discipline of Social Media Intelligence   24
  David Omand, Carl Miller and Jamie Bartlett

3. The Impact of Open Source Intelligence on Cybersecurity   44
  Alastair Paterson and James Chappell

Part II   Open Source Intelligence and Proliferation

4. Armchair Safeguards: The Role of Open Source Intelligence in Nuclear Proliferation Analysis   65
  Christopher Hobbs and Matthew Moran

5. Open Source Intelligence and Proliferation Procurement: Combating Illicit Trade   81
  Daniel Salisbury

Part III   Open Source Intelligence and Humanitarian Crises

6. Positive and Negative Noise in Humanitarian Action: The Open Source Intelligence Dimension   103
  Randolph Kent

7. Human Security Intelligence: Towards a Comprehensive Understanding of Complex Emergencies   123
  Fred Bruls and A. Walter Dorn

Part IV Open Source Intelligence and Counterterrorism

8 Detecting Events from Twitter: Situational Awareness in the Age of Social Media 147
Simon Wibberley and Carl Miller

9 Jihad Online: What Militant Groups Say About Themselves and What It Means for Counterterrorism Strategy 168
John C. Amble

Conclusion 185
Christopher Hobbs, Matthew Moran and Daniel Salisbury

Index 188

Tables and Figures

Tables

6.1	Global information networks: Past, present and future	112
8.1	Highest rated and lowest rated tweets	153

Figures

7.1	Causal pathways of human security	128
7.2	Human security intelligence model	132
8.1	Tweets containing either the first or the last name of an Olympian arriving in the tweet-stream every two minutes	151
8.2	Number of tweets sent every two minutes, between 17:00 and 19:30 on 31 July 2012, expressed as a ratio of the number sent in the previous two minutes	152
8.3	A candidate event: A 'pre-event' and possible 'event' tweet-streams	152
8.4	Tweet-stream on 6 August 2011 between 18:00 and 24:00	156
8.5	Stage 1 signals of some example terms: 'tottenham', 'riot', 'police' and 'lol'	157
8.6a	Original signal	158
8.6b	First-level analysis	158
8.6c	Second-level analysis	158
8.6d	Third-level analysis	159
8.6e	Fourth-level analysis	159
8.6f	Resulting wavelet analysis	159
8.7	Weighted graph showing 'community structure' – similar signals – between terms	161
8.8	Example of clusters drawn out from Figure 8.7	161

Acknowledgements

This book is the product of both a longstanding interest in open source intelligence (OSINT) and a desire to build on the experience and benefits gained from applying OSINT tools and techniques in our research. Furthermore, our work has brought us into contact with a vibrant community of researchers and practitioners who deal with OSINT in various aspects of their work. We were thus presented with an opportunity to bring together the expertise and experiences of colleagues, both at King's College London and elsewhere, with a view to gaining an insight into the ways in which OSINT is understood and employed in different fields of research. By doing this, we hope to offer the reader a snapshot of what is a rapidly growing area of research and activity.

A number of colleagues and friends supported us in this work. We are grateful for the support of colleagues at the Centre for Science and Security Studies, a research centre based in the Department of War Studies at King's College London. Various discussions and exchanges about OSINT and related issues helped to develop our approach to this book. We are also very grateful to the contributors for providing us with original and diverse insights into OSINT, both in terms of its development and current uses, and in terms of its future potential. Molly Berkemeier provided valuable support in the final stages of preparing the manuscript. We would also like to thank the anonymous reviewer at Palgrave for useful comments and suggestions. Staff at Palgrave, in particular Julia Willan, Ellie Davey-Corrigan and Harriet Barker, have also been a pleasure to work with.

Christopher Hobbs, Matthew Moran and Daniel Salisbury

Contributors

John C. Amble is Managing Director of Global Torchlight LLC, a security and risk management consultancy. He has operational experience on three continents as an officer in the US Army, including deployments to both Iraq and Afghanistan. He has also served as an intelligence officer at the Defense Intelligence Agency, as part of the US military's chief counterterrorism intelligence task force. He is currently working towards a PhD at King's College London where his research focuses on regional Islamist insurgent organisations and their evolution as nodes within the transnational jihadist movement.

Jamie Bartlett is the Head of the Violence and Extremism Programme and Director of the Centre for the Analysis of Social Media at Demos. He is the co-author, with Sir David Omand and Carl Miller, of #intelligence – the first framework for the ethical and effective collection of social media intelligence.

Fred Bruls is a Royal Netherlands Air Force reserve major in the Dutch 1 CIMIC Battalion. From August 2009 to February 2010 he was deployed in Afghanistan as information manager in the intelligence branch (G2) of the Dutch-Australian Task Force Uruzgan (TFU-VII). He holds a masters' in defence studies from the Canadian Forces College, where his dissertation was based on the concept of human security intelligence.

James Chappell is co-founder of Digital Shadows and has more than a decade of experience working as a security architect advising the FTSE100 and central government. Prior to setting up Digital Shadows he was Deputy Head of Security at BAE Systems Detica and he remains a member of the CESG Listed Adviser Scheme (CLAS), GCHQ's trusted security advisers to government.

A. Walter Dorn is Professor of Defence Studies at the Royal Military College of Canada and Chair of the Department of Security and International Affairs at the Canadian Forces College. He specialises in arms control, peace/stability operations and international security. One focus is intelligence in United Nations (UN) operations where he benefits from field visits and deployments. He is author of *Keeping Watch: Monitoring Technology and Innovation in UN Peace Operations* (UN University Press, 2011).

Stevyn D. Gibson lectures on concepts of intelligence, security and risk at the UK Defence Academy and Cranfield University. His intelligence experience includes BRIXMIS, analysis for hostage rescue and intelligence briefing to war headquarters. His PhD examined how open source exploitation contributes to the national intelligence function. He is the author of *The Last Mission* (2007) and *Live and Let Spy* (2012).

Christopher Hobbs is Lecturer in Science and Security in the Department of War Studies at King's College London. A physicist by training he has more than five years of experience in applying open source intelligence (OSINT) techniques to nuclear proliferation issues and has developed training courses in this area for the European Defence Agency. His latest book, *Exploring Regional Responses to a Nuclear Iran: Nuclear Dominoes?*, was recently published by Palgrave Macmillan.

Randolph Kent directs the Humanitarian Futures Programme at King's College London. Established at the end of 2005, it is designed to enhance the adaptive and anticipatory capacities of humanitarian organisations to deal with the types of threat that need to be faced in the future. He accepted his present post after completing his assignment as UN Resident and Humanitarian Coordinator for Somalia in April 2002. Prior to his assignment in Somalia, he served as UN Humanitarian Coordinator in Kosovo (1999), UN Humanitarian Coordinator in Rwanda (1994–1995), Chief of the IASC's Inter-Agency Support Unit (1992–1994), Chief of the UN Emergency Unit in Sudan (1989–1991) and Chief of Emergency Prevention and Preparedness in Ethiopia (1987–1989).

Carl Miller is the Research Director of the Centre for the Analysis of Social Media at Demos, and a research associate at the International Centre for Security Analysis at King's College London. His interests lie in creating new ways to learn about people and society from researching social media, and to use these methods to inform policies, decisions and responses to social problems.

Matthew Moran is Lecturer in International Security in the Department of War Studies at King's College London. His research interests include nuclear non-proliferation and the methods and practice of OSINT. His latest book, *Exploring Regional Responses to a Nuclear Iran: Nuclear Dominoes?*, was recently published by Palgrave Macmillan.

Sir David Omand is a visiting professor in the Department of War Studies at King's College London, the former Director of the UK GCHQ and former Permanent Secretary at the Home Office.

Alastair Paterson is co-founder and Chief Executive Officer of Digital Shadows, an OSINT cybersecurity monitoring service. He specialises in designing 'big data' risk and intelligence systems with a particular focus on cybersecurity. Before founding Digital Shadows he was International Propositions Manager at BAE Systems Detica, working primarily with national security clients in Europe, the Gulf and Australasia.

Daniel Salisbury is a researcher at the Centre for Science and Security Studies in the Department of War studies at King's College London, where his work focuses on non-proliferation issues. Specifically, he works on Project Alpha, a UK government-sponsored project which seeks to engage the private sector in non-proliferation and export controls. Prior to his current position he worked at the International Institute for Strategic Studies in London.

Simon Wibberley is a researcher at the Text Analytics Group in the Department of Informatics at the University of Sussex. His research interests lie in statistical text analytics and he specialises in real-time text-stream analysis, event detection and entity recognition. Current projects include developing state-of-the-art event-detection and event-characterisation techniques for use on Twitter.

Abbreviations

ALNAP	Active Learning Network for Accountability and Performance
AMISOM	African Union Mission in Somalia
AOI	area of interest
AOR	area of responsibility
AP	Additional Protocol
API	application programming interface
AQIM	al-Qaeda in the Islamic Maghreb
ASCOPR	areas, structures, capabilities, organizations, people and events
ASIC	all source intelligence cell
BBC	British Broadcasting Corporation
BJP	Bharatiya Janata Party
CCRP	California Coastal Records Project
CIA	Central Intelligence Agency
CIMIC	civil-military cooperation
CISPA	Cyber Intelligence Sharing and Protection Act
COMINT	communications intelligence
COP	common operational picture
COSP	Community Open Source Program
CSA	Comprehensive Safeguards Agreement
DCI	Director of Central Intelligence
DDOS	Distributed Denial of Service Attacks
DF	document frequency
DIME	diplomatic, information, military, economic
DIMEFIL	diplomatic, information, military, economic, financial, intelligence, law enforcement
DNI	Director of National Intelligence
DNS	domain name system
EFP	explosively formed projectiles
ELINT	Electronic Intelligence
EU	European Union
EUU	End User Undertaking
FBIS	Foreign Broadcast Information Service
GIA	Armed Islamic Group
GSMA	GSM Association
GTD	Global Terrorism Database
HSI	human security intelligence
HUMINT	human intelligence

IAEA	International Atomic Energy Agency
ICRC	International Committee of the Red Cross
ICT	information and communication technology
ICU	Islamic Courts Union
IDF	inverse document frequency
IED	improvised explosive device
IFRC	International Federation of Red Cross and Red Crescent Societies
IMU	Islamic Movement of Uzbekistan
IMINT	imagery intelligence
IP	internet protocol
IRC	International Rescue Committee
ISE	Integrated Safeguards Environment
ISI	Islamic State of Iraq
IT	information technology
IVR	interactive voice response
JIC	Joint Intelligence Committee
JMAC	Joint Mission Analysis Centre
JOC	Joint Operations Centre
JTAC	Joint Terrorism Analysis Centre
KLD	Kullback–Leibler Divergence
LINKS	Livestock Information Knowledge System
MASINT	measurement and signature intelligence
MFO	Multilateral Force and Observers
MOD	UK Ministry of Defence
MONUSCO	United Nations Stabilization Mission in the Democratic Republic of the Congo
MRAP	mine-resistant ambush protected
MSF	Medecins Sans Frontieres
MTCR	Missile Technology Control Regime
MUJAO	Movement for Oneness and Jihad in West Africa
NATO	North Atlantic Treaty Organisation
NGO	non-governmental organisations
NLP	natural language processing
NNWS	non-nuclear weapons state
NPoCC	National Police Coordination Centre
NPT	Treaty on the Non-proliferation of Nuclear Weapons
NSG	Nuclear Suppliers Group
NSS	UK National Security Strategy
OPCW	Organisation for the Prohibition of Chemical Weapons
OSC	Open Source Center
OSINT	open source intelligence
OSIS	Open Source Information System
PIRs	priority intelligence requirements

PKO	peacekeeping operation
PMESII	political, military, economic, social, infrastructure and information
PMESII-PT	political, military, economic, social, infrastructure, information, physical environment and time
PO	peace operation
POC	protection of civilians
PSC	private security company
PSO	peace support operation
PTS	Procurement Tracking System
RADINT	radar intelligence
RICC	Regional Information Collection Centre
RIPA	Regulation of Investigatory Powers Act 2000
ROE	Rules of Engagement
SALI	Sustainable Agriculture Livelihoods Innovations initiative
SDSR	Strategic Defence and Security Review
SEG	State Evaluation Group
SEO	search engine optimisation
SGIM	Division of Safeguards Information Management
SGIM-ISF	State Factors Analysis Section in the Safeguards Division of Information Management
SIEL	Standard Individual Export Licence
SIGINT	signals intelligence
SLA	state-level approach
SNA	social network analysis
SOCMINT	social media intelligence
SOPA	Stop Online Piracy Act
SQL	Structured Query Language
START	Study of Terrorism and Responses to Terrorism
TEC	Tsunami Evaluation Coalition
TECHINT	technical intelligence
TF	term frequency
TTA	Trade and Technology Analysis Team
UAV	unmanned aerial vehicle
UN	United Nations
UNAMA	UN Assistance Mission in Afghanistan
UNDP	UN Development Programme
UNFICYP	UN Peacekeeping Force in Cyprus
UNSCR	United Nations Security Council resolution
USD	United States Dollars
WMD	Weapons of Mass Destruction
YHUMINT	young human intelligence

Introduction

Christopher Hobbs, Matthew Moran and Daniel Salisbury

The twenty-first century has seen a revolution in how publicly accessible, or 'open source', information is created, stored and disseminated. Driven by the rapid growth of the Internet and the World Wide Web, as well as the widespread adoption and advancement of mobile communication technology, the use of open sources has permeated the fields of intelligence, politics and business, to name but a few. This revolution has impacted significantly on how people acquire information, express ideas and interact with each other, both socially and professionally. Crucially, while traditional sources and channels of information have made great efforts to adapt to this new virtual environment and retain their presence as gatekeepers of information – many established media sources, for example, now publish large amounts of content exclusively online – the rise of user-generated content, particularly social media, has drastically transformed the information landscape. From the 500 million 'tweets' per day on Twitter, to the 98 million daily blog posts on Tumblr, we are now only a few keystrokes away from a potentially global audience.[1] Moreover, as these tools increase global connectivity, people seem increasingly willing to project their thoughts, opinions and observations into cyberspace. The process of information generation has been opened up to the masses and the sheer quantity of open source information now available online is staggering.

As in other fields, these developments have had a profound effect on the intelligence community. While open source information has long figured in the work of intelligence analysts, it has been conferred with a new status and legitimacy in recent years, moving from the periphery of intelligence efforts to become a core component of analytical products. Indeed, various high-ranking figures in the US intelligence community have for many years claimed that open sources can provide upwards of 80 per cent of intelligence needs – a claim that Stevyn D. Gibson explores in some detail in this volume (Chapter 1). This increased emphasis on open source intelligence (OSINT) – that is to say, the exploitation of open source information for intelligence purposes as part of a broader, all-source intelligence process – has served to provide contextual detail to classified sources which are often limited in scope and fragmented. OSINT can provide background, fill gaps and create links between seemingly unrelated sources, resulting in an altogether more complete intelligence picture. Moreover, due to its open source nature, OSINT can, for the most part, be readily shared and does not present the problems normally associated with the exchange of sensitive information between governments and other organisations.

These changes in the role and perceived value of OSINT are evidenced by the changes that have taken place in the intelligence community. In the US, for example, the establishment of the national Open Source Center (OSC) under the Director of National Intelligence (DNI) in 2005 marked an important milestone.[2] The OSC is an organisation dedicated to the systematic collection and integration of media reports, user-generated online content and any other relevant types of publically available information into the US intelligence cycle. The importance of OSINT in US intelligence efforts was further cemented by the creation of a new managerial position – Assistant Deputy Director of National Intelligence for Open Source – to oversee and coordinate the OSC and, on a larger scale, the growing role played by open sources in the US intelligence enterprise.[3] Moreover, these changes in the US have been reflected to varying degrees in other intelligence communities around the world.

It is not only within the intelligence community that the use of open sources has had wide-ranging implications. The information revolution has affected all fields of research and action. Beyond the efforts of the intelligence community to better integrate OSINT into the all-source intelligence process, many other types of actor are also looking to better integrate open source analysis into their work. From non-governmental organisations (NGOs) to the business community, developments in open source methodologies and practice hold the key to new and valuable insights and analysis.

In practical terms, the current conflict in Syria provides a timely and highly relevant example of the use and value of OSINT. At the time of writing, the Organisation for the Prohibition of Chemical Weapons (OPCW) has begun the process of securing and destroying chemical weapons stockpiles and capabilities declared by the Assad regime.[4] This process is the culmination of months of political and diplomatic activity prompted by allegations of chemical weapons use, most importantly on 21 August 2013 in the suburbs of Damascus. Publically available intelligence assessments produced by the US, France and the UK, among others, claimed that there were significant grounds to believe that the Assad regime had carried out this high-casualty attack on rebel forces. The US intelligence report, for example, was a key pillar supporting the Obama administration's efforts to secure both congressional authorisation and public support for a potential military intervention in Syria, even if the subsequent Russian initiative to convince the regime in

Damascus to commit to giving up its chemical weapons capability meant that military intervention was averted.[5] Similarly, in the UK, a publically available Joint Intelligence Committee (JIC) report presented the case for action, and only defeat in parliament stopped David Cameron's plans to join a potential US-led intervention.

Crucially, these reports relied heavily on evidence gleaned from open sources. The first publicly released intelligence assessment came from the UK JIC on 29 August 2013. This report stated that there were 'no plausible alternative scenarios to regime responsibility', an assessment made with the 'highest possible level of certainty following an exhaustive review by the Joint Intelligence Organisation of intelligence reports plus diplomatic and open sources'.[6] Significantly, the assessment recognised the amount of open source information available on the attack, thus highlighting the value of OSINT in the overall assessment. The following day the White House released a more detailed assessment based on a 'significant body of open source reporting'.[7] The document gave details of the range of sources used to inform the analysis, including 'videos; witness accounts; thousands of social media reports from at least twelve different locations in the Damascus area; journalist accounts; and reports from highly credible nongovernmental organizations'.[8]

These reports and, more importantly, the value that they attributed to open sources in the intelligence process were significant in that they are among the first occasions that the role of OSINT has been so extensively credited in intelligence assessments of such high importance. Of course, this is not to suggest that the emphasis on open sources was completely free of ulterior political motives, or that the value of the open sources used was beyond question. Highlighting the role of open source in the attribution process, for example, provided the relevant governments with a means of diverting attention from the moral and analytical sensitivities associated with covert intelligence – a significant issue in an environment that continues to be overshadowed by the intelligence-related issues that surrounded the 2003 invasion of Iraq. Furthermore, the veracity of the open sources used in these assessments was questioned in the subsequent public debate. For example, commentators asked how videos of the chemical weapons attack could be verified. This question was an important one considering that clear incentives likely existed for elements of the opposition to encourage a Western intervention. On a larger scale, while open sources clearly provided important contextual information, could they provide a 'smoking gun'? Open sources clearly showed the aftermath of a chemical attack. However, could they logically and reliably lead to the conclusion that Assad was responsible?

In general terms the fact that a number of the world's most sophisticated intelligence communities publically highlighted the importance of open sources to their intelligence efforts reflects the growing importance and utility of OSINT. However, the questions raised regarding the role and value of OSINT in the analysis of the Syrian chemical weapons attack touch on some of the enduring issues associated with this rapidly developing area of the intelligence field. On the one hand, then, OSINT presents researchers and analysts with a wealth of opportunities and potential. From the study of online terrorist recruitment to exploring how social media can be used as sources of sociopolitical analysis, OSINT can provide new and exciting data and insights. On the other hand, OSINT poses a number of challenges and obstacles – technical, political and ethical – that must be navigated with care.

In this context, this book takes a fresh look at the subject of OSINT and explores the new approaches, opportunities and challenges that this emergent field offers at the beginning of the twenty-first century. With a focus on three key areas of international security – nuclear proliferation; humanitarian crises; and terrorism – it aims to provide readers with an insight into the latest and most original research being conducted on the subject. The chapters are written by established academics, intelligence specialists, postdoctoral researchers in the early stages of their career, and postgraduate researchers in the final stages of their doctoral work. As a result, the chapters included illustrate the remarkable scope and vitality of research currently being conducted under the broad heading of 'open source intelligence'. The volume's strength lies in both the timeliness of the three security issues themselves and the novel manner in which they are addressed.

The book is presented in four parts. The first considers new methods and approaches in broad, conceptual terms, contextualising some of the new sources, approaches and methodologies which have characterised advances in OSINT in recent years. Stevyn D. Gibson (Chapter 1) begins by exploring the role of OSINT and broadly defines its value to the intelligence function. He challenges popular assumptions regarding both the capabilities and the limitations of OSINT and argues that cultural, organisational and ideological contexts exert an important influence on OSINT and must be taken into consideration in attempts to assess the value of OSINT.

David Omand, Carl Miller and Jamie Bartlett (Chapter 2) introduce the concept of social media intelligence (SOCMINT) as a branch of OSINT. They argue that the analysis of social media offers the possibility of new levels of social, political and ideological insight, and claims that the advances made in data analytics methodologies make social media analysis of immense value, both to the intelligence community and beyond. Alastair Paterson and James Chappell close Part I (Chapter 3) by exploring the impact of OSINT on cybersecurity. They describe the dangers that the availability of open source information about businesses in their digital presence poses to information assets and business activities in an increasingly web-based society. They go on to explore some innovative ways of mitigating these risks.

The three subsequent parts build on the concepts and issues raised in the more general opening section, addressing OSINT's relevance and application to three topical issues in international security: nuclear proliferation, terrorism and humanitarian crises.

In Part II on OSINT and proliferation, Christopher Hobbs and Matthew Moran (Chapter 4) begin by exploring the value of OSINT in assessing states' nuclear intentions and capabilities, focusing on the approach of the International Atomic Energy Agency (IAEA). From political statements to scientific and technical publications, open source information can provide a range of clues regarding a state's nuclear trajectory. This is followed by Daniel Salisbury (Chapter 5), who considers the opportunities and challenges that OSINT provides in understanding how states illicitly procure technologies for their nuclear and missile programmes. Using the growth in publically available information about illicit procurement as a starting point, he discusses the value of largely untapped information held by the private sector, and conversely the role that OSINT can play in informing industry about the risks posed by present-day illicit procurement attempts by states such as Iran and North Korea.

Part III explores OSINT in the context of humanitarian crises. Randolph Kent (Chapter 6) begins by exploring the growing reliance on social media as a means of dealing with humanitarian crises. While acknowledging and detailing the benefits of social media to those working to mitigate the effects of humanitarian crises, he also examines the drawbacks of this new aspect of OSINT. He argues that 'negative noise' (contradictions and inconsistencies in information) can add confusion to humanitarian operations, and he proposes systemic approaches to mitigate this problem and promote greater reliability and authenticity. Fred Bruls and A. Walter Dorn (Chapter 7) argue that a new concept, human security intelligence (HSI), holds the key to the comprehensive understanding of humanitarian crises that is essential for field operations to be a success. Based on the concept of 'human security', Dorn and Bruls argue that the idea of HSI derives, to a large extent, from the power and value of OSINT.

Part IV considers the value of OSINT in terms of understanding terrorism. It begins with Carl Miller and Simon Wibberley (Chapter 8) who build on the theme of social media set out in Part I. They explore the ways in which social media can be harnessed to detect events and improve responses to large-scale emergency situations, such as terrorist attacks. Moving the focus from the response to terrorist attacks to the groups themselves, the John C. Amble (Chapter 9) presents an analysis of jihadist groups' online presence. He argues that the notion of a global jihadist movement is both reductive and limiting, particularly in terms of counterterrorism strategy. Amble draws on the media releases of three terrorist groups – al-Qaeda in the Arabian Peninsula, Lashkar-e-Taiba and Boko Haram – with a view to illustrating the range of identities, beliefs and ideologies that exist within the jihadist movement. Ultimately, the chapter argues that OSINT offers a means of gaining a more nuanced insight into the individual identities of terrorist groups and that this approach should form the basis of distinct, tailored approaches to mitigating the threat posed by a particular group.

In general terms, the focused and subject area-specific chapters highlighting the uses, benefits and challenges of OSINT in particular security contexts complement the more conceptual chapters set out in Part I to provide readers with a comprehensive and far-reaching analysis of an area that has grown in importance over the past two decades.

Notes

1. See Richard Holt, 'Twitter in Numbers', *The Telegraph*, 21 March 2013, http://www.telegraph.co.uk/technology/twitter/9945505/Twitter-in-numbers.html; and 'Press Information', Tumblr website, http://www.tumblr.com/press.
2. 'INTellingence: Open Source Intelligence', CIA website, 23 July 2010 (updated 30 April 2013), https://www.cia.gov/news-information/featured-story-archive/2010-featured-story-archive/open source-intelligence.html.
3. 'ODNI Announces Establishment of Open Source Center', ODNI News Release No. 6–05, 8 November 2005, https://www.fas.org/irp/news/2005/11/odni110805.html.
4. Julian Borger, 'Syria: Chemical Weapons Inspectors Begin Securing Assad Regime's Arsenal', *The Guardian*, 3 October 2013.
5. 'Obama to Seek Congress Vote on Syria Military Action', *BBC News*, 1 September 2013.
6. 'Syria: Reported Chemical Weapons Use – Letter from the Chairman of the Joint Intelligence Committee', *UK Cabinet Office*, 29 August 2013, https://www.gov.uk/government/uploads/system/uploads/attachment_data/file/235094/Jp_115_JD_PM_Syria_Reported_Chemical_Weapon_Use_with_annex.pdf.
7. 'Government Assessment of the Syrian Government's Use of Chemical Weapons on August 21, 2013', *Office of the Press Secretary, The White House*, 30 August 2013, http://www.whitehouse.gov/the-press-office/2013/08/30/government-assessment-syrian-government-s-use-chemical-weapons-august-21.
8. Ibid.

Part I

Open Source Intelligence: New Methods and Approaches

1
Exploring the Role and Value of Open Source Intelligence

Stevyn D. Gibson

> A proper analysis of the intelligence obtainable by these overt, normal and aboveboard means would supply us with over 80 percent, I should estimate, of the information required for the guidance of our national policy.
>
> *Allen Dulles*

OSINT and the mythology of the '80 percent rule'

Allen Dulles's testimony to the Senate Committee on Armed Services on 25 April 1947 was only nine pages long and was hastily written[1] but, in it, he began the process of the demystification of the art of intelligence, adding:

> Because of its glamour and mystery, overemphasis is generally placed on what is called secret intelligence, namely the intelligence that is obtained by secret means and by secret agents... In time of peace the bulk of intelligence can be obtained through overt channels, through our diplomatic and consular missions, and our military, naval and air attachés in the normal and proper course of their work. It can also be obtained through the world press, the radio, and through the many thousands of Americans, business and professional men and American residents of foreign countries, who are naturally and normally brought in touch with what is going on in those countries.

Dulles's 80 per cent claim represented the first notable attempt to quantify the open source contribution to intelligence and has since been widely repeated by practitioners, commentators and customers alike. The North Atlantic Treaty Organisation (NATO) considers that the exploitation of OSINT provides 80 per cent of the final product for arms control and arms proliferation issues.[2] For his part, Hulnick suggests that 80 per cent of US Cold War analysis could have been taken from open sources.[3] Steele's view is that, across the board, OSINT can provide 80 per cent of what any government needs to know and 90 per cent for private sector organisations.[4] For some collection agencies, 90 per cent is the norm.[5] The Central Intelligence Agency's (CIA's) Bin Laden unit, for example, noted that '90 percent of what you need to know comes from open source intelligence'.[6] EUROPOL goes further, suggesting that the contribution might be as high as 95 per cent for counterterrorism issues.[7] Similarly, the 1996 Aspin-Brown Commission remarked that 'In some areas... it is estimated that as much as 95 per cent of the information utilised now comes from open sources'.[8] Responding to the report of the 1997 US Commission on Secrecy, the grand-master of US foreign policy, George F. Kennan, wrote to US Commission Chairman Senator Daniel P. Moynihan[9]:

> It is my conviction, based on some 70 years of experience, first as a Government official and then in the past 45 years as a historian, that the need by our government for secret intelligence about affairs elsewhere in the world has been vastly overrated. I would say that something upward of 95 per cent of what we need to know could be very well obtained by the careful and competent study of perfectly legitimate sources of information open and available to us in the rich library and archival holdings of this country. Much of the remainder, if it could not be found here (and there is very little of it that could not), could easily be nonsecretively elicited from similar sources abroad.

Finally, but not exhaustively, in December 2005, former Deputy Assistant Director of Central Intelligence W.M. Nolte stated that 95–98 per cent of all information handled by the US intelligence community derives from open sources.[10] It would seem, then, that those best placed to judge regard OSINT as constituting a significant majority of the intelligence effort.

Regardless of the general attribution and subjective estimate of OSINT's efficiency, the obvious question ought to be: what is it 80 per cent or 95 per cent of? Is it output measured by paragraphs in a final intelligence report, actions enumerated by arrests and threat interdictions, or clearly observable policy achievements?

Thus, this oft-quoted estimate that 80 per cent or more of final intelligence product is generated from open source exploitation is a mischievous 'red herring'. How is the figure calculated? What is the yardstick of measurement? Where are the repeatable, corroborable data by which it is determined? The evidence is anecdotal; subjectively assessed rather than methodically derived. Furthermore, while Dulles and Kennan suggest that 80 per cent of what the intelligence function needs to know can be found from open source information, this is not quite the same as saying that 80 per cent of what the intelligence function 'knows' is OSINT. The former is a claim for the usefulness of OSINT as a deliberate discipline in itself; the latter is a verdict on the limitations of traditional secret sources.

Moreover, this broad 80 per cent estimate does not equate to all intelligence subjects equally, or simultaneously. In the mid-1990s the US Government's Community Open Source Program (COSP) estimated the OSINT contribution to be in the range of 40 per cent overall, while specific contributions, depending upon target difficulty, ranged from 10 per cent in denied-area, secret-issue matters to 90 per cent in international economics. This COSP estimate may be the only methodically derived data point for the evaluation of OSINT's contribution. As Markowitz suggests, much of the chatter surrounding the 80 per cent claim might be no more than circular reporting of Dulles's original estimate. Once stated by respected members of the intelligence community it passes into lore.

Therefore these 'estimates' of OSINT's contribution to final intelligence product, even if correct, are merely expressions of efficiency – inputs related to outputs. This may form the basis of an important argument about the allocation of scarce intelligence resources. Although, in terms of those scarce resources devoted to OSINT, this 80 per cent label neither carries the weight of an 800 lb gorilla nor enhances its 'second-class' status relative to closed INTs. Most importantly it does not reveal any understanding of OSINT's contribution to decision- or policy-making effectiveness.

In this context, it would seem more useful to explore the effectiveness of OSINT – what it can do for the intelligence function. In this regard, it can replicate 'secret' sources, form the matrix to bind all other intelligence sources together and still has its own distinct attributes to offer. Yet OSINT is no more a 'silver bullet' for policy than closed sources.

Some 'INTs' are more equal than others

Regardless of the '80 percent rule', the proportion of resources devoted to OSINT is nowhere near comparable to closed. OSINT may be formally and deliberately exploited within the intelligence community, but it is regarded as less equal than others. This is due to a number of cultural barriers that have somehow come to shape OSINT's potential.

Many myths and misconceptions serve to confound OSINT's contribution to intelligence: that it is in competition with secret intelligence, rather than complementary to, if not thoroughly enmeshed with, closed; that it resides solely on the Internet rather than in magnetic, film, paper and other non-digital sources; that it is exclusively text-based and in English rather than also oral, image-based, and in many languages; that it is conducted overtly, when collectors may hide their interest at a conference, mask their intentions in the academic papers that they deliver or 'anonymise' their IP address when interrogating websites; that it is exclusive to the public sector, when, by definition, it is available to many with a cause, including the private, academic and other non-governmental sectors; that it is free to collect or assess, rather than requiring specialised effort and increasingly expensive effort as greater value is added; that the greatest added value may come from any sector – private sector product is not 'inferior' to public sector product, nor OSINT necessarily 'inferior' to closed 'INTs'; that it is excused the usual 'rules' of information-working commonly applied to construct assessment in support of decision-making, rather than be validated for accuracy, relevance and timeliness in the same way that journalism and research should be; and, not least, that OSINT cannot provide a 'smoking gun', when many historical examinations of 'intelligence surprise' show those surprises being pre-trailed in the press, and countless examples of contemporary social networking media 'confessions' demonstrate that much evidence is already in plain sight in these media.[11]

Thus the concern with effectiveness is important in one key regard – it should prioritise, or at least influence, the treatment of OSINT within agencies and across national intelligence machineries. All of these cultural, organisational and technical misconceptions underline the necessity for a distinct OSINT tradecraft, appropriate tools and techniques, specialised software and equipment, a 'familiarity' with contemporary information and communication technology (ICT) and a befitting budget.[12] The establishment of the Open Source Center (OSC) in the US goes part way towards realising such a vision; but, as Bean observes, the OSC remains inside a closed environment and subject to high-level office politics.[13]

Any obsession should be with the meaning rather than the number. As RAND and Gill and Phythian have all noted, intelligence effectiveness is a slippery concept to pin down, let alone measure.[14] Odom similarly bemoans the fact that nowhere within the intelligence community are inputs related to outputs:

> Because the DCI has never made the effort to impose a similar system (to the Defense Department) on resource management in the Intelligence Community, its consolidated Intelligence Community budget does not effectively relate inputs to outputs.[15]

Odom argues that instead of traditional benchmarks of quantity and quality of data gathered, effectiveness should be measured by how much output is used by, and meets the needs of, its customers. Interestingly, the US Army begins to construct such an argument: 'determining whether PIRs [priority intelligence requirements] have been answered'.[16] The US Joint Chiefs expand it by recognising that intelligence evaluation is undertaken by the customer, based upon 'the attributes of good intelligence: anticipatory, timely, accurate, usable, complete, relevant, objective, available'.[17]

The UK intelligence community similarly confuses efficiency with effectiveness. Its Intelligence and Security Committee relates inputs to outputs through 'top-level management tools' ensuring that 'business' objectives are met within the intelligence agencies, the resulting public service agreements and service delivery agreements reflecting a transfer to resource-based accounting processes. Worse, they equate process with purpose and confuse means with ends.

Thus today's 'best' determination of effectiveness seems to reside with the customer or the accountant. Yet customer satisfaction, business targets and balancing scorecards generate little meaningful insight into the effectiveness of intelligence in relation to policy objectives. In order to evaluate OSINT's contribution, it seems crucial to understand how it is effective, both absolutely within the intelligence function and relative to closed intelligence.

Ecclesiastes Chapter 1, Verse 9

> The thing that hath been, it *is that* which shall be;
> and that which is done *is* that which shall be done:
> and *there is* no new *thing* under the sun.
> Standard *King James Bible*

In 1808, Wellington assembled his generals before departing for the Peninsular War and admonished them for their ignorance of Napoleon's new French infantry formations being openly reported in *The Times*. In 1826, Henry Brougham, the radical Whig politician, established the Society for the Diffusion of Useful Knowledge. Its aim was 'to impart useful information to all classes of the community'.[18] This utopian dream of knowledge transforming society sounds strikingly similar to the present-day goal of Google.[19] The society closed in 1848. Its product was considered erratic and miscellaneous; one might add idealist, naïve and absent of political purpose. It thought that the provision of information was the end game and in doing so elevated means to ends. Contemporary open source evangelists, such as those behind the now infamous WikiLeaks, pioneer similar utopian visions for 'open' information. Finally, during the nineteenth century, the commercial open source publication *Jane's Fighting Ships* was established in 1898.

In March 2002, John Darwin canoed out into the North Sea from the English seaside town of Seaton Carew and faked his suicide. By February 2003 he had moved back into the family home to live secretly with his wife, Anne. In March 2003 a death certificate was granted and his wife began the fraudulent process of recovering the insurance on her husband's supposed death. Between 2003 and 2007 the couple travelled abroad and constructed a new life away from the UK authorities as well as a few suspicious people at home. They set up a home and business in Panama. In December 2007, John Darwin walked into a London police station claiming amnesia;

he was 'reviving' his identity in order to satisfy new Panamanian investment laws. The police had already begun investigations into possible fraudulent activity three months earlier. Yet they failed to connect Anne Darwin's frequent trips abroad with her husband's adventures in Panama. Five days after his return, the *Daily Mirror* published a photograph of the Darwins with their Panamanian estate agent. The photograph had been taken in 2006 and published on the Internet as part of the estate agent's marketing campaign. This photograph had been discovered by a 'suspicious' member of the public, who had simply typed 'John', 'Anne' and 'Panama' into Google. She informed Cleveland police of the result. The police expressed 'surprise' at the simplicity of this public-spirited piece of detective work but it took the *Daily Mirror* to convince them. The information had been openly available in plain sight for 18 months.

Hindsight is a wonderful thing – the scourge of intelligence practitioners. There are myriad examples of information in support of defence, foreign policy or law enforcement being hidden in plain sight. Today, such is the proliferation of social networking that law enforcement, security and intelligence agencies exploit these mobile, digital, Internet-based resources for evidence of law-breaking, self-incrimination and acts preparatory to law-breaking. Whether they do this routinely, efficiently and effectively, let alone ethically, and without subverting privacy or justice, remain important questions.

The formation of institutional 'state-sponsored' organisations by which open sources of information are exploited for 'modern' or 'industrial' intelligence purposes can be pegged to the creation of the UK's BBC Monitoring Service in 1938 (now BBC Monitoring).[20] Its US equivalent – Foreign Broadcast Monitoring Service, later Foreign Broadcast Information Service (FBIS), and today's OSC – emerged in 1941.[21] Both were formed in response to the invention of radio – in particular, its use in the 1930s as a tool by the Axis Powers for the dissemination of propaganda. Not only did they monitor broadcast media as a collection activity in its own right but they also gauged the response to our own propaganda broadcast as part of wider and more nuanced information operations – much as they do now.

That 'monitoring' became categorised as 'open' and 'interception' as 'closed', was partly a reflection of the intended nature of the data transmitted – public versus secret – and partly to conceal from the target that it was being collected against. As the two world wars merged into the Cold War, technical intelligence (TECHINT) and signals intelligence (SIGINT) penetrated the secretive nature of 'Eastern Bloc' society; and, the satellite platform bequeathed to collection what the Internet is today. 'At home', secrecy dominated the capabilities-oriented intelligence requirements of the Cold War in order to protect sources, methods and product from 'the enemy'. Although OSINT played the junior partner to closed collection, organisations such as the Soviet Studies Research Centre in the UK evolved alongside the BBC Monitoring Service to reveal secrets – capabilities – and to chip away at mysteries – intentions – through examination of open source media.

Thus the exploitation of open source information is not new. Its contemporary prevalence is a reflection of the increasing volume, immediacy and accessibility of today's mobile digital ICT. Today, contemporary ICT, built on digitisation and miniaturisation, and enhanced by the mobile phone, provides new platforms for intelligence, as well as a 'new', more 'open' globalised society. OSINT is reblooming as a result of both this technological evolution and the 'opening up' of formerly closed societies.

Like the concept of intelligence, the definition of open source exploitation – the exploitation of information legally available in the public domain – is debatable. What is the public domain? How do we determine 'legally available'? Who does the exploitation? When WikiLeaks dumped thousands of secret documents into the public domain via the Internet in 2010–2011, the US Library of Congress, bizarrely, was forbidden from using them in its own assessments because they remained classified. Yet they were available to anyone with an Internet connection, including, presumably, Library of Congress researchers on PCs at home.

What OSINT can do...

Compared with the more traditional or esoteric intelligence techniques, it is often faster, more economical, more prolific, or more authoritative.

Herman L. Croom

In 1969, Croom summarised much of the contemporary debate surrounding OSINT in just seven pages and encapsulated it in the quote above.[22] He recognised the key benefit of OSINT's contribution. He constructed a case for its efficacy using nuclear weapons proliferation as well as the developing international situations of Africa, Latin America and Southeast Asia as casestudies. He argued for a more equitable policy towards intelligence resource allocation. He recommended the establishment of an open source agency – *outside* the CIA – specifically instructed to treat this intelligence species.[23] His vision did not materialise.

More recently, Sands set out five contributing factors that OSINT offers relative to closed sources of intelligence: an assessment frame of reference; the protection of closed material; credibility; ready access; and enhanced assessment methodology.[24]

(Stevyn) Gibson's research across contemporary intelligence, security, law enforcement and corporate organisations establishes seven high-order factors describing the contribution of OSINT – its effectiveness – to the wider intelligence function[25]:

1. *Context.* Without exception all intelligence agencies recognise that open sources of information represent a 'matrix' in which to conduct their work.

16 OSINT: New Methods and Approaches

Described variously as 'a first port of call', 'stocking filler', 'background' or simply 'contextual material', it represents the most widely acclaimed attribute of OSINT.[26]

2. *Utility.* This is a synonym for speed, volume and cost. Intelligence practitioners – public and private – recognise that it is quicker, more productive and cheaper to collect open sources of information before closed. It is simply more immediately useful to analysts. The volume, and the immediate availability of that volume, may prove to be a 'game-changing' aspect of information-working, although, it is worth remembering that much collected information is never processed or analysed; there are comparably sophisticated tools to process those increased volumes; and these challenges are not new but merely new to us. Dealing with so-called 'big data' will be as much a reflection of us – attitude and technology – as the 'bigness' of the data.

3. *Benchmark.* Closed single-source intelligence collection agencies use OSINT as a benchmark against which their sources are gauged. They recognise that it is uneconomic and unprofessional to disseminate intelligence product derived from closed sources when it is publicly available. Moreover, benchmarking is used in both ways – closed information can also be used to challenge open source information. Confusingly, some single-source agencies now claim difficulty in distinguishing open from closed sources.

4. *Surge.* Clandestine intelligence, particularly human intelligence (HUMINT), is not an activity that can be turned on and off like a tap. Conversely, open sources, already 'out there', are more easily 'surged' than traditional sources. They provide an holistic all-source capability: satellite imagery is available commercially off the shelf; 'news' can be aggregated, searched and sorted on the Internet; 'citizen journalists' or 'bloggers' help to unravel the mysteries of 'uncertainty'; while the sharing of intelligence via an 'Intelipedia' moves analysis towards a real-time product.[27] When 'surprise' happens – kidnap to revolution – analysts and policy-makers turn to OSINT for a 'first cut'.

5. *Focus.* Open sources can be used to both direct further collection efforts and provide the illusive here and now of 'point' information. Thus, focus implies both direction (targeting onto new 'leads') and acuity (high granularity).

6. *Communicability.* Practitioners claim that information derived from open sources is easier to share and disseminate. Security concerns shrink to who collects it rather than where it originates from. Yet the dominant culture of secrecy confounds this theoretical patency.

7. *Analysis.* This is designed to achieve three things: situational awareness; an understanding of situations; and some tentative forecasting effort. Broadly, this comprises at least five varieties of product: current intelligence; database

or knowledge creation; forecasts; warnings and indicators; and 'red teaming'. Each of these is achievable and undertaken by OSINT. However, the degree to which open source collectors undertake analysis varies considerably. In private information brokerages, analysis is undertaken by individuals or groups comfortably interchanging between all functions of the intelligence cycle in order to create product. In public sector agencies, analysts and collectors are broadly separated. Interestingly, today, BBC Monitoring acknowledges that it formally conducts an analysis of its own collection. Similarly, open source procedures vary between 'push' or 'pull' systems. In some agencies a 'push' system operates whereby open source expertise sits (literally) alongside an analyst or team of analysts. In others a 'pull' system operates where analysts have to go to centralised open source cells with their requirements and wait for a response.

Notwithstanding the value of these high-order factors, they still only represent what OSINT can do for the intelligence function. They are no evaluation of effectiveness advancing the ideological or political objectives of decision-makers that intelligence supports. Equally, such relative contributing factors have still to be incorporated into any meaningful exploitation of open sources by 'Western' national intelligence functions encumbered by a proclivity towards closed and classified information-working.

Contemporary information-working and the implications of so-called 'big data'

History has witnessed several significant transformations in ICT: the invention of the alphabet to enable writing; the cipher to aid mathematics; the printing press to democratise communication; electricity, the telegraph, and transistors to deliver instant communication; steam and internal combustion engines to deliver the messenger; and chips with everything. Notably, the printing press contributed to two revolutionary periods – the Renaissance and the Reformation. Contemporary ICT – particularly the combination of mobile phone and Internet – has similarly contributed in recent decades to 'revolution', change and the democratisation of decision-making. Certainly, the democratisation of information is increasing if one 'counts' the number of people being 'connected' to the Internet and the amount of information available there.

There are some cautionary notes. First, the availability of more information in the public sphere may confer a quantitative rise but it does not infer any similar qualitative improvement.[28] One of the 'complaints' of contemporary information-working – characterised by metaphors such as information society, digital age, big data and new social media – is the cliché of 'drinking at a fire-hydrant'. Yet if information is your business, where else would you want to drink? If you cannot apply the necessary filters, funnels, adapters and pumps to control the flow then you may be consigned to sup with a straw from obscure puddles. Presumably there is little desire to return to 'small data' – whatever that was. Yet, more importantly, rather than blame intelligence analysis on either too much or too little information, it would be more useful if analysts – and their policy-masters in particular – interrogate their underlying assumptions about the way the world is, as well as recognise that their underpinning ideological frameworks are too absent or weak to take their societies forward. Open source exploitation will not do this; critical thinking might.

Second, the increased immediacy of information does not excuse the necessary activity of analysis. The oft-cited '90 per cent of everything is crap' rule also pertains. Nowhere does this seem more apparent than in the 140-character confines of the 'Twitosphere'. Third, regardless of the medium of communication, communicators still need to have something meaningful to say. In the political sphere, while Twitter, Facebook, SMS messaging or the Gutenberg printing press can contribute significantly to political processes, the primacy of any political movement remains the strength of its motivating idea. Moreover, the rules of information-working apply as much to these new sources as they do to traditional sources. Thus *sapientia* is as likely (or not) to come from open source exploitation as it is from the exploitation of secret sources.[29]

The 'because we can generation', or why is Lord McAlpine trending[30]

While the achievement of strategic or operational objectives more meaningfully reflects the effectiveness of both open and closed exploitation, the contribution of a wider democratisation of information to optimal decision-making remains to be seen.

In 430 BC, Pericles of Athens suggested that 'Although only a few may originate policy, we are all able to judge it.' Unsurprisingly, and in the light of contemporary ICT, this maxim still holds. Arguably, today's communication systems afford a greater degree of visibility than the athenaeums of Ancient Greece. We can see almost everything going on and often in near real time. Yet similar maxims also hold. As Onora O'Neill points out in her 2002 Reith lectures, no amount of 'enforced' openness and transparency regimes will guarantee an increase in honesty or a decrease in deception; people will simply reveal what they want you to see.[31] Thus the need for closed or secret intelligence methods remains extant.

However, unlike closed methods where techniques, tools or agents, if not breaking foreign laws to retrieve vital information, pose difficult moral choices between the needs of security versus the transgression of civil liberties such as privacy, in open source exploitation the moral hazard seems less clear. We exploit open sources because we can. In the public domain, any moral transgression seems unclear if not invisible. By the same token, we are content to place what once might have been considered private information into the public domain for most of us to 'look at'. Of course, this does not reflect the capabilities of ICT; rather, it reflects a broader contemporary disregard for privacy and a narcissistic culture of exhibitionism. As Jurgen Habermas notes in his 'paradox of the public sphere', although the barriers to entry into public debate are being dismantled, the points of entry and entrants themselves are increasing, or the immediacy and cost of entrance improving, this has resulted neither in any commensurate improvement in reflection by its participants nor in any significant improvement in the quality of debate.[32]

The moral hazard does not reside within open or closed exploitation; it resides in the changed attitudes of contemporary Western society. As Brendan O'Neil points out, the open, immediate and voluminous incontinence of the 'public sphere' is not a technical fault of the extraordinary connectivity conferred by the Internet and the World Wide Web. It is a cultural erosion of the fine line between private thought and public behaviour.[33] It is the logical conclusion of those individuals who are so disconnected from society, absorbed with themselves and incapable of selfless restraint that they expose themselves publicly and 'blame' new technology rather than a lamentable decline in longstanding critical thinking. So there ought to be another maxim for all forms of media and people who use it: 'because we can does not mean we should'. The moral obligation to think whether 'we should' is not removed or excused by the fact that 'we can'. This choice is not a new one; perhaps it is simply a harder one.

The revolution will not be televised

This famous Gil Scott-Heron lyric might be applied to all ICT, including contemporary forms of social media. Thus, in Scott-Heron's terms, the 'revolution' is also very unlikely to be the product of Twitter, Facebook or YouTube. Certainly, the revolution may be downloaded or uploaded by these tools and technologies. Indeed, these tools may even contribute to coordinating the revolution. For example, in 2013 the Free Syrian Army called down artillery fire on Assad government forces using Twitter. Although this seems 'revolutionary', it is no more than British Royal Marines using satellite communications in 1982 or the Royal Navy exploiting Marconi's wireless telegraphy in 1899. Revolutions will be sparked by an idea constructed in the human mind and driven by a collective, coercing and irresistible political will to win arguments. It is this enduring nature of politics that will form the crucible for ideological direction and leadership. The conduct of politics, like warfare, will change with and be changed by prevailing technologies of the day.

New forms of communication – new social media – are not of themselves politically threatening. Much of Twitter's 140-character content, for example, offers little existential challenge to a well-established and confident political structure and system. It may be the playground for the vain, the stupid, the bored or the simply indifferent, but it reflects no political 'revolution'. On the contrary, its remoteness, superficiality and 'virtuality' – absent of real human interaction – present a stupefying anaesthetic to genuine political activity that the inhabitants of real political power, rather than fret over, should welcome unless, of course, they are absorbed by it themselves.

It is not ideological conviction that manifests itself across the Facebook 'Twitosphere' but a shallow, selfish narcissism that detracts from notions of solidarity in any real and meaningful political sense. That we can communicate more easily, more voluminously and more instantly between place 'A' and place 'B' does not equate to real change in either place. This is not to say that new forms of social media are not useful targets at the operational level, especially for solving or preventing criminal activity; reassuring the public; engaging the digital public in cooperative crime-solving; and countering 'rumour' on the social media street. Whether it warrants a new title – SOCMINT – is discussed elsewhere in this book. Arguably it could be called young human intelligence (YHUMINT) for now.[34]

Thus the purpose of intelligence – support to decision-making, speaking truth to power or determining who knows what in sufficient time to make use of it – remains extant. By contrast, the conduct of intelligence – how it is done – is continually evolving, commensurate with changes in ICT. The increasing availability, immediacy and volume of open sources generated by the latest transformation in ICT are reflected in the formal and deliberate acceptance of OSINT as a new information class alongside HUMINT or TECHINT. Certainly, today, open sources remain recognised and exploited across the intelligence, security and law-enforcement communities.

Plus ça change...

Open source exploitation – whether it is publicly available 140-character Tweets, North Korean journals or commercial satellite imagery – is but one contributing strand to the intelligence function and information-working more generally. Like all information-working disciplines it is subject to methodical norms and rules that all information-working disciplines are subject to – validity, accuracy, reliability and corroborability at least.

OSINT's specific contribution to intelligence has been categorised here into seven high-order factors: context; utility; benchmark; focus; surge; communicability; and analysis. These qualitative factors – measures of efficacy perhaps – are useful in directing the intelligence function in three important ways: the allocation of resource; the focus for training, policy and doctrine; and the efficient tasking of the intelligence community.

Yet the paradoxical location of OSINT inside the established closed environment, its weak resource allocation, and the stubborn culture of intelligence practitioners and customers – 'classification-obsession', bureaucratic resistance to change, 'office politics' – hamper OSINT's full integration, appropriate weighting and status.

The extent of OSINT's effectiveness in relation to outcomes beyond the confines of the intelligence function – ideological purpose, the national interest and foreign policy – remains illusory. Like intelligence more broadly, measures of effectiveness remain rooted in perceived added value to the customer, accounting procedures, and pseudoscientific scorecards rather than related to political objectives. Indeed, these subjective self-affirming yardsticks hinder any linking of the intelligence function to political, ideological or economic goals.

Equally, it is important to acknowledge that OSINT is not the be-all and end-all of intelligence. There will remain a need to discover 'facts' that others do not wish to be revealed as long as international relations, crime and commerce remain 'competitive sports'. OSINT may be a key source for the intelligence function in today's globalised, open and increasingly digitally inclined world, but it is no 'silver bullet' for decision-makers, contrary to the beliefs of some advocates. Closed and secret intelligence efforts remain necessary.

The greater challenge remains – as always – not in finding 'solutions' to politics within a 'better' intelligence function but in the intelligence function recognising external shortcomings of weak politics. Globalisation, mobile digital communications and the democratisation of information render a deeper contextual understanding, visible to others beyond just intelligence, security and law-enforcement communities. The traditional 'predictors' seem to be unable to detect readily apparent concerns: a post-Cold War ideological vacuum; a risk-management decision-making framework substituting for proper political leadership; and the contemporary ascendancy of the pursuit of security over the pursuit of liberty.[35] Policy-makers should know something more of the world beyond intelligence – certainly beyond the world of secret intelligence – and intelligence practitioners should recognise the limitations of contemporary Western polities, derailed from the pursuit of Enlightenment objectives. Whether intelligence communities can deliver this or decision-makers recognise the deficit remain debatable issues.

Thus the challenges for the total information business, of which intelligence and, more specifically, OSINT are parts, are not simply how to deal with the increasing volume of information, the sharing of classified sources, or honestly balancing open and closed capabilities. These can be dealt with procedurally by technological means and culturally by attitudinal shifts. They are matters of operational concern. The greater challenges for the intelligence community reside in presenting a genuine picture of reality to decision-makers; regaining confidence in judging rather than measuring intelligence effectiveness; and reconnecting power and society to purpose in the sense of what we are for, rather than deferring to process and what we are against.

Notes

1. Peter Grose, *Gentleman Spy: The Life of Allen Dulles* (Boston: Houghton Mifflin, 1994), pp.525–528. Citing the US Senate Committee on Armed Services, Hearings on the National Defense Establishment, 1st Session, 1947.
2. NATO, *Open Source Intelligence Reader* (Norfolk, VA: SACLANT Intelligence Branch, 2002).
3. Arther S. Hulnick, *Keeping Us Safe: Secret Intelligence and Homeland Security* (Westport, CT: Praeger, 2004), p.6.
4. Correspondence, Robert David Steele – Gibson, 3 October 2003.
5. NATO, EUROPOL, EU, UK MOD, Swedish MOD, Dutch MOD, US DIA, CIA, UK HMRC-LE to list a few in the public/government sector.
6. Susan B. Glasser, 'Probing Galaxies of Data for Nuggets', *The Washington Post*, 25 November 2005, http://www.washingtonpost.com/wp-dyn/content/article/2005/11/24/AR2005112400848.html.
7. Stevyn D. Gibson, 'Open Source Intelligence (OSINT): A Contemporary Intelligence Lifeline' (Cranfield University: PhD thesis, 2007), Appendices: Interview EUROPOL 4, https://dspace.lib.cranfield.ac.uk/handle/1826/6524.
8. US (Brown-Aspin) Commission on the Roles and Capabilities of the United States Intelligence Community, Preparing for the 21st Century: An Appraisal of US Intelligence (Washington DC: US Government Printing Office, 1996).
9. George Kennan wrote to Senator Daniel Moynihan in preparation for the senator's testimony to the Committee on Governmental Affairs, United States Senate Hearing on Government Secrecy, 7 May 1997.
10. William M. Nolte (former Deputy Assistant Director of Central Intelligence for Analysis and Production), presentation to the Oxford Intelligence Group, 5 December 2005.
11. See also Stephen Mercado, 'A Venerable Source in a New Era: Sailing the Sea of OSINT in the Information Age', *Studies in Intelligence: Journal of the American Intelligence Professional* (2004), Vol. 48, No. 3, pp.45–55.
12. 'Familiarity' with contemporary ICT is more likely to be found in the under-50s. See Mercyhurst College, Department of Intelligence Studies, http://intel.mercyhurst.edu/.
13. Hamilton Bean, *No More Secrets: Open Source Information and the Reshaping of U.S. Intelligence* (Westport, CT: Praeger Security International, 2011).
14. RAND, *National Security Research Division & Office of DNI, Towards a Theory of Intelligence* (Washington: RAND, 2006), pp.26–29; P. Gill and M. Phythian, *Intelligence in an Insecure World* (Cambridge: Polity Press, 2006), p.18.
15. William Odom, *Fixing Intelligence for a More Secure America* (New Haven: Yale, 2003), p.32.
16. US Army, US Army Field Manual – Intelligence 2–0 (2004), pp.1–8.
17. US Department of Defense Joint Chiefs of Staff, Joint and National Intelligence Support to Military Operations JP 2–01 (2004), III, pp.56–57.
18. See 'Society for the Diffusion of Useful Knowledge Papers', AIM25 archive database website, http://tinyurl.com/c39d8po.
19. 'To organize the world's information and make it universally accessible and useful'. See 'Company Overview', Google website, http://www.google.com/about/company/.
20. For a comprehensive history, see O. Renier and V. Rubinstein, *Assigned to Listen: The Evesham Experience 1939–1943* (London: BBC Books, 1986).

21. For a concise history, see S. Mercado, 'Open Source Intelligence from the Airwaves: FBIS Against the Axis, 1941–1945', *Studies in Intelligence* (2001), Fall–Winter, p.11.
22. Herman L. Croom, 'The Exploitation of Foreign Open Sources', *CIA Historical Review Program* (1969), Vol. 13, No. 2, p.131.
23. Ibid., pp.129–136.
24. Amy Sands, 'Integrating Open Sources into Transnational Threat Assessments', in Jennifer Sims, Burton Gerber (eds.), *Transforming US Intelligence* (Washington, DC: Georgetown University Press, 2005), pp.63–78.
25. Stevyn D. Gibson, 'Open Source Intelligence (OSINT)'.
26. Rolington's description of President Clinton's intelligence needs succinctly demonstrates the significance of context. Alfred Rolington, 'Objective Intelligence or Plausible Denial: An Open Source Review of Intelligence Method and Process since 9/11', *Intelligence and National Security* (2006), Vol. 21, No. 5, pp.741–742.
27. Intellipedia is the US response to Wikipedia for intelligence practitioners.
28. For a comprehensive treatment of the so-called 'information age', see Frank Webster, *Theories of the Information Society, 3rd Edition* (Abingdon: Routledge, 2006).
29. Stevyn D. Gibson, 'Future Roles of the UK Intelligence System', *Review of International Studies* (2009), Vol. 35, No. 4, pp.917–928.
30. This refers to unfounded allegations made against Lord McAlpine in 2012 – complete with the icon;-) – by Sally Bercow on her Twitter account.
31. Onora O'Neil, *A Question of Trust: The BBC Reith Lectures 2002* (Cambridge: Cambridge University Press, 2002).
32. Frank Webster, *Theories of the Information Society*, pp.263–267.
33. Brendan O'Neil, 'Don't Blame the Web for Our Urge to Blab', *The Telegraph*, 30 July 2013, http://tinyurl.com/k7cxr6x.
34. Jamie Bartlett, Carl Miller, Jeremy Crump and Lynne Middleton, *Policing in an Information Age* (London: DEMOS, 2013).
35. Stevyn D. Gibson, *Live and Let Spy* (Stroud: The History Press, 2012), pp.204–221.

2
Towards the Discipline of Social Media Intelligence

David Omand, Carl Miller and Jamie Bartlett

Open source SOCMINT today: A practice but not a discipline

We are living through a revolution in how we communicate. Every month, 1.2 billion of us now use Internet sites, apps, blogs and fora to post, share and view content.[1] Loosely grouped as new, 'social' media, these platforms provide the means by which the Internet is increasingly being used: to participate, to create and to share information about ourselves and our friends, our likes and dislikes, movements, thoughts and transactions. The largest, Facebook, has over a billion regular users, but the linguistic, cultural and functional reach of social media is much broader, from social bookmarking to niche networks, video aggregation and social curation.[2] LinkedIn, a specialist business network, has 200 million users, the Russian-language VK network 190 million users and the Chinese QQ network 700 million users.[3]

All of these interactions leave a digital trace. Social media is now the largest body of information about people and society we have ever had. Analysis of these new social spaces is already widespread throughout the private sector for marketing and brand management, in politics for election planning and as a burgeoning topic of academic interest.

This form of analysis is now making the leap into the public sector. The August 2011 disorder in England was a wake-up call regarding the power of social media, and how unable the police and other public agencies were to understand or harness it.[4] In 2012 we wrote #intelligence, arguing that the police and intelligence services have a responsibility to react and adapt to the explosive rise of social media.[5] It can be harnessed by the public sector as a source of intelligence – something we dubbed SOCMINT.

Since 2012 it has become steadily clearer how, around the world, police and security agencies are seeking to acquire SOCMINT capabilities. In the UK it is now an accepted form of intelligence on social disorder on the part of the Home Office. An 'all sources hub' – which includes open source social media – was established by the Metropolitan Police to collate intelligence to help with the policing of protests at the time of the London Olympics. Most

UK police constabularies already use social media as a basis for investigation, albeit in a limited way, or as a way of seeking information from the public and communicating with the public – for example, during an emergency or following a major accident.[6]

Whilst SOCMINT has emerged as a practice, it is not yet a coherent academic discipline or distinctive intelligence tradecraft. It is a scattering of isolated islands of emphasis rather than a united body of learning, method or example. Towards this emergence, this chapter aims to describe its emerging contours: to codify the capabilities that have emerged and the opportunities that they have created, and the risks and hurdles that they must commonly face and surmount – methodological, legal and ethical – in order to usefully contribute to decisions in a way that is publicly supported.

The emergence of a coherent body of pooled learning and experience on the use of SOCMINT is vital. The fundamental purpose of intelligence – including SOCMINT – has been defined as being to improve the quality of decision-making by reducing ignorance.[7] With its varied use and broad interest, the analysis of social media sources can produce intelligence in a very large number of things, capable of improving decision-making in a very large number of organisations in both the private and the public sectors. For the public sector, it can decisively contribute to public safety and security by improving the information available to those who have to make difficult decisions, often under acute time pressure. It is capable of doing so on three different levels.

The first level, and most fundamental, is by helping the authorities establish situational awareness as events unfold. From analysis of social media, essential facts about the world can be identified: 'when' events occurred, 'what' is happening now, 'where' and involving 'whom'.

The second and more complex task is to construct the best possible causal explanation of events as they have been observed, the 'why' and 'how' of the activities in question, from terrorism to street crime. This second level of analysis demands a very high degree of general understanding of the phenomenon in question and the terms in which participants habitually express themselves.

Given adequate explanations of the events and the motivations of the participants then, it may be reasonable to look for a prediction of how events will unfold. This third level of analysis involves the most complex use of intelligence, including that derived from social media, to answer the inevitable questioning from the political and operational authorities about 'what next' and 'where next'. Essential for planning a law-enforcement or security response is this ability to take explanation and prediction, and to attempt to model the situation – for example, how it might unfold under different forms and scales of response by the authorities addressing the 'what if'. The access that social media intelligence can give to near real-time developments in a fast-moving situation, and the geographical granularity of information, has the potential to transform the sensitivity of the response, thus minimising the risks of alienating the community or provoking by too great a show of force the very violence that the response is intended to defuse.

Challenges to effective SOCMINT

SOCMINT does, therefore, have considerable potential to improve all three levels of analysis and thus contribute directly to public security. However, its potential has yet to be realised due to a focal, crucial problem: its use, and the people who use it, are disparate, disaggregated and diverse. It is conducted across the public, private and academic sectors, and it spans disciplines from the computer sciences and ethnography to advertising and brand management. With aims ranging from understanding the topography of social networks comprising millions of individuals to the deep, textured knowledge of the social worlds of individuals and small groups, the techniques that underlie SOCMINT reflect specific disciplinary traditions, aims and assumptions, and rarely converge into a pooled body of approaches.

Crucially, these diverse and often isolated practices face a body of common challenges that stand in the way of their ability to generate reliable and ethical contributions to decision-making in ways that enjoy public confidence and support. The first challenge is efficacy. Digital social spaces are new sites of social activity with new norms, language uses, and psychological and social influences, and therefore they carry new meanings and significances. The tools and techniques that are used to understand these spaces are also new, and commonly struggle to meet the standards of evidence that are needed; produce datasets that are representative, interpret them in ways that credibly and meaningfully expose an underlying social reality, and validate these analyses in the face of the many deceptions that we know to be widespread in social media. There is consequently a heightened risk of getting things wrong, either as false positives or negatives – of either spotting a phenomenon that doesn't exist or missing one that did.

The second challenge is to be able to conform to publicly accepted ethical standards. As the potential for the use of SOCMINT becomes recognised by security authorities around the world, and as the ambitions of law-enforcement and intelligence agencies for the use of SOCMINT become better known, a wide debate has opened up about the potential for misuse of these techniques both by the private sector and by state authorities. Like all forms of intelligence, its use by the authorities can have real-world consequences for individual citizens. There is consequently a wide demand for a publicly argued, consistently applied framework of regulation and oversight. Yet many of the laws around the world that enshrine and guarantee citizens' protection from abusive SOCMINT were passed before social media existed.

Ways of learning about people and society benefit from shared bodies of expertise – doctrines, ethics, wisdoms and methods – and experts who can generate new insights. Indeed, the validating stipulations of standards and thresholds that they create define good practice, and ensure that efforts made along those lines inform, not misinform, decisions. Few would argue that the practice of history can be safely pursued without a sound knowledge of historiography, learning how to test the authenticity and validity of source documents and the motivation behind their appearance or conservation. Likewise, the tradecraft of human intelligence relies upon experience in validating sources and helping the users of intelligence by caveating and classifying them – for example, as regular and reliable or as new sources on trial.

The understanding of how SOCMINT can be reliably and ethically used is incomplete and immature. This is an important area for security studies, for we can anticipate that the use of SOCMINT will race ahead of the full understanding of its characteristics. Meeting these common challenges that SOCMINT users face requires the development of the commonalities that it currently lacks: a coherent body of knowledge, expertise, skills, and authorities and communities, and therefore the emergence of SOCMINT as both a distinctive tradecraft and an academic discipline, with signature approaches and standards of collection, analysis, verification, use and regulation.

Some SOCMINT is secret intelligence: it involves overcoming the obstacles that others have put in the way, such as encryption of social media use, restriction to only those authorities approved for access to social media pages, or the setting of privacy controls that have to be overcome to access the raw material. In this chapter we deal only with the use of 'open source' SOCMINT. In this we follow the US intelligence community definition of open source as information that anyone can lawfully obtain by request, purchase or observation.[8] We do not consider either the techniques or the legal considerations needed to legitimately acquire or use closed or private information carried on social media.

Techniques, methods and opportunities

The 'state of the art' in methods and techniques for generating insight from open source social media is rapidly developing on a number of, often isolated, fronts. We describe those techniques, methods and opportunities that we regard as most important to this nascent discipline's ability to contribute to each of these three analytical layers, and explain the challenges that they together face in doing so. We recognise that others of course exist and that still more will exist in the future.

SOCMINT to inform security and law enforcement

We look first at how SOCMINT can establish and generate evidence and more generally contribute to situational awareness: 'what' has happened, involving 'whom' and 'when'.

As evidence

Social media are already fora where criminals associate, communicate and conspire to commit offences, and where offences themselves – including new crimes particular to social media – are committed. Facebook has been used to try to hire hitmen, groom the targets of paedophiles, steal identities and fatally cyberbully victims.[9] By 1999, nearly all known terrorist groups had established a presence on the Internet.[10] It is still unclear how terrorist and extremist groups will end up exploiting the opportunities created by social media, but a picture is emerging that social media use is significant in a number of terrorism cases in the UK.[11] The use by criminal organisations and gangs more generally of the Internet and social media is expected to increase.[12]

One obvious opportunity that SOCMINT offers law enforcement is to assist in the prosecution of criminals by obtaining evidence from the public. In its most direct form, platforms such as Twitter can act as a new channel of communication – similar to a hotline – allowing the public to share information, including videos and photos, in real time with the police. Analysis conducted by Demos found that of 20,000 tweets sent to the Metropolitan Police between 17 and 24 May, some 20 per cent were referrals of evidence – including eyewitness accounts of offline crimes, and evidence of allegedly criminal activity on Twitter itself. The most common kinds of complaint or referral made to the police fell into the broad family of hate speech: alleged to be racist, anti-Semitic, threatening or inciting violence. Alongside hate speech, tweets were identified alleging driving infractions, fraud, involvement in riots, paedophilia, child abuse, drug-taking, cyberbullying and animal abuse.[13] It has to be recognised that such social media use is not always tactically welcome to the police, and in particular the use by media outlets of social media sites can make the initial investigations by the police harder by prompting a deluge of irrelevant or even malicious misinformation. But, on balance, police services see the social media dimension as positive.

'SOCMINT as evidence' is increasingly a proactive practice of modern policing to encourage the reporting of committed crime and public suspicions of criminal activity that can cue further investigation by traditional methods. In 2009, Strathclyde Police launched Operation Access, which used social networking sites such as Facebook to uncover criminal activity by identifying weapons carriers – resulting in 400 people being questioned.[14] It is also used as a general investigatory tool, to identify missing persons and identify individuals. Internationally, Canada has used social media to collect successfully valuable evidence on criminal activity such as illegal gun ownership.[15] The New York Police Department launched a new unit with the primary responsibility of tracking criminals on Facebook, Myspace, Twitter and other social networking sites.[16]

As situational awareness

Social media can also be collectively analysed to spot and characterise 'events': whether political, cultural, commercial or in case of emergency. These events could be intrinsic to social media, such as a particular type of conversation or trend, or indicators or proxies of events that have occurred offline.[17] During the 2011 Egyptian revolution – for instance, 32,000 new groups were formed and 14,000 new pages created on Facebook in Egypt.[18]

Across many platforms, social media users play a number of different roles in exchanging information that can detect events. They can generate information about events first-hand. They can request information about events. They can 'broker' information by responding to information requests, checking information and adding more information from other sources, and they can propagate (and amplify or distort) information that already exists within the social media stream.

Event detection SOCMINT (as Chapter 8 in this volume illustrates) is increasingly using algorithmic statistical modelling to identify and characterise events by observing profiles of word or phrase usage on social media over time. It attempts in particular to pinpoint anomalous spikes of certain words and phrases together as a signal that an event may be occurring. This has clearest practical utility for intelligence in contributing to situational awareness of rapidly developing and chaotic events – especially emergencies.

The platform of most significant interest for event detection to date is Twitter. In addition, however, an area of crucial development has been combining different social media information, both textual and non-textual, across different platforms. One study used YouTube, Flickr and Facebook, including pictures, user-provided annotations and automatically generated information, to detect events and identify their type and scale.[19] Finding ways of reliably assessing the 'temperature' of a group through linguistic analysis of social media usage is a high priority for current academic research with several promising avenues being explored.

SOCMINT to aid understanding and explanation

As observed earlier, SOCMINT can contribute towards a second level of intelligence analysis aimed at understanding the dynamics of the phenomenon under study, including the motivations of those taking part in social media usage, adding the 'why' and 'what for' to the 'what, where, when and who' of situational awareness.

As 'big data' insight

The world is being 'datafied' – observed, measured, recorded and analysed – at previously unprecedented scales. This has led to leaps in human comprehension across a number of fields, exposing relationships, dynamics, processes and information about causes and consequences that were previously unseen.[20] The transfer of people's lives onto digital-social spaces has 'datafied' social life – huge, naturalistic and constantly refreshing bodies of behavioural evidence that are inherently amenable to collection and analysis.[21]

New opportunities to gain insight into social life are opened up by the possibility of subjecting these very large social media datasets to a burgeoning suite of 'big data' analytics tools that can identify and quantify the variables that contribute most to explaining the variances in the observed phenomena. But correlation is not causation, and as much care is needed in social media analysis not to fall into that trap – and the presentation of 'false positives' – as is the case for conventional multivariate statistics.

One of the most promising of these 'big data analysis' tools is natural language processing (NLP). This is a particular application of artificial intelligence and linguistics to allow automated computational systems to derive meaning from the human (or 'natural') language that exists in social media. It typically does this using supervised machine-learning algorithms that produce models based on statistical correlations found in examples provided by human analysts. These models are then able to automatically, and at great scale and speed, classify information about social media into a number of different categories based on their meaning. It is, essentially, a window into datasets that are too large to read.

The most significant application of NLP to social media to date has been to discern 'sentiment', where NLP classifiers sort tweets into broad 'positive', 'neutral' and 'negative' categories. This is widely used in the commercial field to track how brands are mentioned on social media, in order to measure the impact of marketing campaigns, for instance. It is driven by the belief that analysis of these large, constantly updating datasets compliments and improves upon conventional attitudinal research – interviewing, traditional polling or focus grouping.[22]

Three additional and emerging applications of NLP – beyond the recognition of positive and negative sentiment – are significant.[23] The first is to classify tweets into more sociologically informed and informative categories than positive, negative and neutral: such as urgent, calm, violent or peaceful.[24] This has been used to measure whether Twitter reflects or (see below) pre-empts violence or disorder offline. The second is to use NLP classifiers to assess whether data are relevant or irrelevant, rather than positive or negative.[25] This form of relevancy filtering is sometimes known as 'disambiguation'.[26] The third, and very experimental, is to create layers of multiple NLP classifiers to make architectures that are capable of making more sophisticated distinctions.

To understand social networks

Social network analysis (SNA) is an approach that combines sociological and mathematical methods to describe the nature – intensity and frequency – of relationships between 'individual elements', typically people or small groups. It is predicated on the insight that people are powerfully influenced in a number of important psychological and behavioural ways by the social ties that surround them. By establishing the structural characteristics of these social ties, SNA attempts to explain the behaviour of the network overall, and the behaviour of its members based on their location within it, including highly networked 'influencers', specific communities within wider networks, and the 'gatekeepers' between them.

In order to derive SOCMINT, SNA can be conducted on different types of dataset of online activities, including blogs, news stories, discussion boards and social media sites. It attempts to measure and understand those 'network links' both explicitly and implicitly created by the features of the platform, and how the platform is used. These include formal members of particular movements; followers of Twitter feeds; members of forums; communities of interests; and interactions between users. Once the data is gathered it can be used for a number of purposes, ranging from estimating how many individuals are engaged in a specific activity online to understanding the flow of information and influence in complex systems.[27]

Typical SNA on social media includes:

- tracking the increase in content produced about a specific issue or location;
- tracking the spread of a specific piece of information;
- tracking the sharing of information between individuals;
- understanding the complex structures created by the behaviour of individuals that influences the information that other users receive, and the behaviours that those communities adopt.

These analyses have been used to understand how strong and loose networks influence behaviour.[28] This has been applied especially to understand why and how people are influenced (either ideologically or behaviourally) on social media. This has often been driven by a desire to market and reach 'key influencers' within a particular field.[29] For security authorities, potential uses could include better understanding of public behaviours in a crisis, the phenomenon of 'herding' and that of panic buying when shortages of key commodities or services occur. It is also used to understand people's information environments through understanding the relationships that they leverage to acquire information and knowledge.[30]

SOCMINT as an aid to prediction

Broadly, there is a growing sense that the 'big data' revolution – the improvement in the ability of humans to make measurements relating to the world, record, store and analyse them in unprecedented quantities – is making possible a level of accuracy in predictions about many aspects of the world that were previously impossible.[31]

Big data prediction relies on an array of techniques (including sentiment analysis) to produce statistical models that allow what is measured and perceived on social media to be used as a predictive signal for other behaviour, including 'offline' behaviours in the real world.[32] Sparked by an influential paper by Hal Varian, Google's chief economist in 2009, there has been an interest in applying predictive analytics to social media datasets to predict a range of social behaviours and phenomena, from election results to box office numbers and stock market trends.[33] Work on predictive SOCMINT is being carried out in a number of areas, most significantly politics, health and crime.

Political prediction and influence

Correlations of social media sentiment can predict voting decisions. Barack Obama's 2008 (and 2012) presidential campaign used data from Twitter and Facebook to predict which people were strong influencers of the swing voters, and targeted them, not the swing voters themselves.[34] Research from Tweetminster during the 2010 UK general election found that the volume of mentions on Twitter – at the national but not the candidate level – is associated with overall election results.[35] A similar study was undertaken in the German federal election of 2009 (albeit with less success).[36]

Public health

One area that has received a lot of attention is the use of Twitter data to understand the spread of infectious disease for the purpose of public health monitoring. A 2012 paper found that, based on an analysis of 2.5 million geotagged tweets, online ties to an infected person increased the likelihood of infection, particularly where there is a noted colocation (for the obvious reason that they are likely to be living near or together, or to meet).[37]

Crime

Most of the work that has been done on criminal incident prediction relies primarily on historical crime records, and geospatial and demographic information sources. Indeed, a series of recent reports about 'predictive policing' are not based on social media datasets but use existing crime data and other datasets to predict crime hot spots.[38]

However, the possible statistical link between Twitter and crime and disorder is an active research topic, and new tools and methods are currently under development by interdisciplinary research teams.[39] The use of automated language recognition to spot certain types of 'risky' behaviour or criminal intent is also a developing field. Some linguists argue that certain structural, underlying features of sentence and syntax are broadly correlated with general behaviour types, such as anger and frustration, which are subconscious but revealed in the structure of a text, rather than the words used.[40]

When assessing this capability, it is wise to bear in mind Nate Silver's observation that 'prediction in the era of big data is not going very well'.[41] He suggests that humans have a propensity for finding random patterns in noise. With big data, the amount of noise increases relative to the amount of signal, resulting in enormous datasets and producing lots of correlative patterns that are ultimately neither casual nor valuable. We believe that predictive SOCMINT will prove increasingly useful for security purposes, although to date it has rarely been tested in the real world. Most studies have been based on a 'retrospective fit' – that is to say, post-event analysis of pre-event data, which allows researchers time and modelling in order to test various hypotheses.

Challenges to UK SOCMINT

The different categories of SOCMINT mentioned in this chapter are highly disparate in what they aim to do and how they try to do it. However, all of them are potentially open to common challenges that stand in the way of general acceptance of their ability to contribute sensibly to decision-making. These criticisms can be broadly broken down into challenges to credibility and challenges to use.

Credibility of open source SOCMINT

As with any intelligence, there is a pervasive concern about the quality and credibility of information being collected and analysed. Any usable intelligence that will have real-world consequences must meet certain criteria regarding how it is gathered, evidenced, corroborated, verified, understood and applied. The key challenge facing SOCMINT is that, as a new and emerging form of intelligence, there is not yet the human and technological infrastructure to convincingly answer a number of key challenges to its credibility.

Representivity

Much open SOCMINT relies on the analysis of large bodies of aggregated data. These data are often only descriptively useful if they represent an identifiable constituency – either online or offline. This allows the analysis – say a predictive model or an attitudinal insight – conducted on this dataset to be generalised to the population that it represents, whether a specific group, a community or the entire population. However, current conventional sampling strategies on social media construct handcrafted or incidental samples using inclusion criteria that are arbitrarily derived and noted for their inferential weakness.[42]

Veracity

In the course of any analysis, 'measurement veracity' is the assurance that what is being measured is what is described as being measured. Does, for instance, a piece of research on attitudes towards immigration use research instruments – surveys, questionnaires or polls – that assuredly tap into these attitudes? Measuring social media, especially in very large quantities, is a young and rapidly developing field that cannot have recourse to the depth of literature and experience of other academic methodologies, use cases or research contexts. In addition, many of the analytical techniques that are used (especially NLP) are inherently probabilistic. They therefore often produce results and insights that, inevitably, have a measured chance of being incorrect, and may require new formats of caveating and nuancing in order to be appropriately understood and used, as well as recognition that SOCMINT can provide leads for further investigation by more traditional means.

Genuineness

A threat to validity faced by all social research is the measurement effect. This is where the very act of measuring changes the measurement itself, either because those being measured are directly asked something, or more generally are aware of being observed.[43] These measurement effects are often the product of unconscious biases on the part of research subjects.[44]

As with any form of intelligence, the analyst must be alive to the possibility of being deliberately deceived. The openness and anonymity of social media, especially, make them a suitable medium for deceptive tactics. A deliberate intent to mislead could be expressed through the presence of misinformation: fake 'honeypot' accounts and impersonation.[45] These deceptive measures are conducted for a number of reasons, from hoaxes and campaigns, to crime, spam, fraud and gaming. The phenomenon is complex, and while it is difficult to establish overall estimates for how widespread it is as a practice, it is estimated to be common. Facebook recently revealed that seven per cent of its overall users are fakes and dupes.[46,47]

It is difficult to provenance, authenticate and therefore detect, track and predict deceptive practice. Attempts to detect detection dynamically interact with counter attempts to evade such detection. This 'arms race' of sorts continues to be an important area of research.[48] A core aspect of any SOCMINT capability will be the ability, both analyst-led and automated, to weed out false and misleading information.

Reality

The intent, motivation, social signification, denotation and connotation – the meaning – of any utterance is dependent on the context of situation and culture. So far, SOCMINT capability has been dominated by the growth of very powerful quantitative methods. This has led to a state of the art that is on the whole good at counting instances on social media and can indeed do so at truly incredible, unthinkably large scales. However, our ability to count things on social media has outpaced our understanding of what these things mean as social and cultural practices – as symbols, as language games, as rituals, as products of digital worlds ruled by new norms and subjective truths. Without knowing these things, we may misunderstand what we are counting. Overall, to be useful, many 'big data' findings on social media need to be wrapped around a digital ethnography and digital sociology that give their numbers meaning.

Substantiation

In the face of these stiff challenges to credibility, the validation of SOCMINT is especially important. However, there have not yet been developed and tested strategies that are commonly used across SOCMINT practice or social media research to validate whatever is produced.

In many respects, validating any SOCMINT acquired from a single source is much the same as in any other area of investigation, and would require the same standards and methods applied in any human intelligence source: track record, known capabilities, motivations and so on. This validation process requires a common reporting framework that rates the 'confidence' in any piece of freestanding piece of SOCMINT. By pointing out potential vulnerabilities and biases in the acquisition and analysis of the information, we may gauge the importance of the information collected and caveat the conclusions that may be drawn.

However, SOCMINT must be able to be related to other kinds of evidence, to produce an overall picture – the equivalent of an 'all-source assessment'. Social media outputs can be cross-referenced and compared with more methodologically mature forms of offline research. These can be already existing 'gold standard' administered and curated datasets (such as census data, and other sets held by the Office of National Statistics), especially the increasing body of 'open data' that now exists on a number of different issues, from crime and health to public attitudes, finances and transport, or bespoke research conducted in parallel with SOCMINT projects. The comparisons – whether as overlays, correlations or simply reporting that can be read side by side – can be used to contextualise the robustness of findings from social media research. An example would be the overlay of people tweeting about respiratory problems with data on pollution, traffic congestion and air quality.

UK use of open source SOCMINT

As with any other form of intelligence reporting, the ability of SOCMINT to contribute to decision-making depends on it getting to the right people in time, securely, and presented in a format that makes sense to strategic and operational decision-makers, as well as those at the frontline.

To be effective, therefore, SOCMINT must slot into appropriate existing intelligence channels at the strategic, operational and tactical levels. Thus, for example, the outcome of SOCMINT analysis of longer-term trends in crime, domestic social exclusion and extremism will be of interest to Home Office criminologists and policy-makers, and to police planning staffs, and similarly information about social attitudes in relation to countries at risk of instability will be of interest to the assessments staff who serve the Joint Intelligence Committee. Operational analysis from social media of jihadist messaging[49] and its reflection in social networking would be a natural subject for the UK Joint Terrorism Analysis Centre. Tactical near-real-time information about reactions to outbreaks of disorder would be valuable to responding police services and the UK National Police Coordination Centre (NPoCC). For operational and tactical use, specific training will be needed on the nature of SOCMINT and its risks as well as advantages – for example, for strategic and tactical police commanders and additional training for frontline officers who could benefit from the daily use of such intelligence. The value of SOCMINT relative to other forms of intelligence must be evaluated (including closed source SOCMINT obtained under the appropriate legal authority) and the ways in which various types of intelligence can be used in tandem needs to be investigated. The crucial points here are the exact application of SOCMINT in a given circumstance and its 'strength' in relation to other forms of intelligence.

SOCMINT's dissemination must reflect its complexity, scale and dynamism. Social media datasets tend to be both very large and highly disparate, containing both very valuable and totally useless (or even counterproductive) pieces of information. As the problems of validation indicate, valuable pieces of SOCMINT are not immediately indistinguishable from many, many pieces of misinformation, propaganda, lazy half-truths and deliberate falsehoods. Because sorting the wheat from the chaff is therefore not easy, it is difficult to 'triage' SOCMINT – to be able to rapidly prioritise those pieces of information which are more valuable, and act on them despite the many SOCMINT outputs that are also competing for attention. Any disseminated data need to be presented with new procedures and caveats. Depending on the purpose of the SOCMINT, this may range from a footnoted, caveated and in-depth strategic analysis paper, to the operational use of single-screen, real-time visualisations of data available on portable devices.[50]

Legitimacy and law

Intelligence collection has to be conducted according to the law. For many countries, however, the laws that govern intelligence work were passed before social media existed as either a technology or a widespread social practice, and it is not always clear how they are applied. The European Convention on Human Rights enshrines and recognises the value of two fundamental public goods: first, the right to life and the security of the person (article 1, as a basic condition that other rights flow from) and the right to privacy (article 8). Public acceptance of any form of data collection is sustained by it being carried out within clear, publicly argued frameworks that balance these two fundamental values.

For UK public bodies, this trade-off is principally managed by the Regulation of Investigatory Powers Act 2000 (RIPA). Although structured to distinguish between the different legal mechanisms authorising interception of communications (Part 1 of RIPA) and other forms of investigation (Part 2), RIPA is animated by the common precept that the more possibly intrusive the access, the fewer agencies should be authorised to access and use the information, the narrower an acceptable justification for such access should be, and the tougher the oversight of the process has to be. This is based on the crucial associations of proportionality, necessity and legitimate aim.[51]

The problem is that RIPA was passed into law in 2000 before the advent of social media, and only now is guidance being prepared to make clear where authorisation under RIPA might be required in respect of social media. Part of the difficulty stems from the way in which RIPA is constructed. It is intentionally flexible in that authorisation requests are determined on individual cases, involving a proportionality test. Proportionality is partly decided on the basis of measuring whether personal information might be revealed in the course of an investigation; and the expectation of privacy a person might have in relation to the information being collected about them. That is why RIPA authorisation is sometimes required where there is a likelihood that 'private information' will be obtained, even if it comes from a public source.

In our view it is vital that the way in which RIPA – and any other relevant legislation such as the Police Act – is interpreted and used in respect of SOCMINT is publicly debated and agreed, and then codified in publicly available guidance. In other work on the subject we have argued that open source SOCMINT seems most likely to fall within the ambit of activity regulated by Part 2 of RIPA2000 when it amounts to directed (non-intrusive) covert surveillance.

Under current RIPA guidance, while there is a reduced expectation of privacy in public places, covert surveillance of a person's activity in public may still result in the obtaining of private information. As such, some SOCMINT activities, especially those carried out over a sustained period of time to establish 'patterns of life', will legitimately fall under directed covert surveillance because (a) much private information is shared on public domain social media spaces; (b) reasonable expectations of privacy vary; and (c) SOCMINT often uses listening-in technology, such as scrapers, which, while they might not be 'calculated' to be covert, are usually designed in a way to be unseen. Therefore consideration needs to be given to the conditions under which it is decided that online data collection regarding an individual amounts to directed covert surveillance.

At the heart of any decisions should rest the principle of 'reasonable expectation' of privacy that a social media user might have in respect of their information, at both the individual and the societal level.[52] Guidance should also include the length of time during which the authorities may store the social media data that they collect since that is also a matter that requires a balancing of the interests of security (the ability to go back in time after some current event to study its origins) and of privacy.

The reason it is important to have public understanding of how RIPA will be employed is because the methods used to protect society rest ultimately on some form of public acceptability and involvement. Britain's National Security Strategy recognises that security and intelligence work in general is predicated not only on the public's consent and understanding but also on the active partnership and participation of people and communities. Serious erosion of the state's ability to protect its citizens is liable to occur when the state's efforts are not accepted or trusted.[53]

Conclusion: A principled way forward

Open source SOCMINT must therefore be seen to pass two tests. First, it must be seen to have a reasonable prospect of contributing towards public safety by delivering usable results. If it did not, there would be no moral, or indeed financial, argument for it to be collected or used. Second, it must be seen to be legitimate, and enjoy public consent for the broad circumstances regarding how and when it may be used.

As in so many areas of human activity, a balance has to be struck between competing public goods; in the case of SOCMINT, public safety and security on the one hand and the public's legitimate right to privacy of normal life on the other. We believe that maintaining the balance rests on the ability of those authorities that wish to access, analyse and use publicly available social media activity to be able to satisfy a number of ethical tests. We suggest here five such principles, adapted from those earlier suggested by Sir David Omand for the intelligence community more generally in his book, *Securing the State*.[54]

Principle 1: there must be sufficient, sustainable cause. This first and overarching principle forces the big picture to be taken into account: the overall purposes that could justify the acquisition by a public body of capabilities to gather, understand and use social media data. There is a danger that a series of SOCMINT measures – each in themselves justifiable – taken together will be regarded as comprising an unacceptable capability for overall surveillance, with the consequent loss of confidence in a medium that is of obvious intrinsic value beyond security. In terms of allowing wide use of open source surveillance therefore just because it can be done does not mean that it should be done.

Principle 2: there must be integrity of motive. This principle refers to the need for integrity throughout the whole intelligence system, from the statement of justification for access, accessing the information itself, through to the objective analysis, assessment and honest presentation of the resulting intelligence.

Principle 3: the methods used must be proportionate and necessary. There is a well-established principle in law enforcement that the extent of harm that could arise from any specific action being taken should be proportionate to the harm that it is being sought to prevent.

Principle 4: there must be right authority, validated by external oversight. There is a general principle that there must be an audit trail for the authorisation of actions that may carry moral hazard with an unambiguously accountable authority within a clear chain of command, allowing individuals to seek redress in cases where abuse of power is suspected. This is an important way in which proportionality and accountability are tested in practice.

Principle 5: recourse to intrusive intelligence, including some open source SOCMINT, must be a last resort if more open sources can be used. Because of the moral hazards of all intrusive secret intelligence-gathering methods, those authorising such operations should ask whether the information could reasonably be expected to be obtained through other means, ranging from fully open sources and methods to information freely volunteered from within the community.

SOCMINT faces common ethical and methodological barriers that require, we argue, a common response. As islands of SOCMINT practice and development have now emerged, it is now time for them to converge into both a tradecraft and a discipline, with the expertise and experts necessary to ensure, through informing, understanding and predicting, that they can reliably contribute towards important decisions. Only through this can its true potential be unlocked as a new way of keeping society safe, consistent with society's own values and principles.

Notes

1. 'With 1.2 Billion Users Worldwide, Social Networking Usage Patterns Vary by Country and Region', *Emarketer*, 29 February 2012, http://www.emarketer.com/Article/Where-World-Hottest-Social-Networking-Countries/1008870.

2. For a visual sense of the entire span of social media platforms, see B Solis and JESS3, 'The Conversation Prism: The Art of Listening, Learning and Sharing', www.theconversationprism.com.
3. Craig Smith, 'How Many People Use the Top Social Media Apps and Services?', *Digital Marketing Ramblings*, 23 June 2013, http://expandedramblings.com/index.php/resource-how-many-people-use-the-top-social-media/.
4. As the subsequent review noted: Her Majesty's Inspector of Constabulary, *The Rules of Engagement: A View of the August 2011 Disorders*, 2011, http://www.hmic.gov.uk/media/a-review-of-the-august-2011-disorders-20111220.pdf.
5. David Omand, Jamie Bartlett and Carl Miller, *#Intelligence* (London: Demos, 2012).
6. Jamie Bartlett, Carl Miller, Jeremy Crump and Lynne Middleton, *Policing in an Information Age* (London: Demos, 2013), p.4.
7. David Omand, *Securing the State* (London: Hurst & Co, 2010).
8. National Open Source Enterprise, *US Intelligence Community Directive 301*, 11 July 2006 (Washington DC: DNI), https://www.fas.org/irp/dni/icd/icd-301.pdf.
9. '20 Infamous Crimes Committed and Solved on Facebook', *Criminal Justice Degrees Guide*, 1 March 2012, http://mashable.com/2012/03/01/facebook-crimes/.
10. See Ines Von Behr, Anaïs Reding, Charlie Edwards and Luke Gribbon, *Radicalisation in the Digital Era: The use of the internet in 15 cases of terrorism and extremism* (Santa Monica, CA: RAND, 2013).
11. Aaron Zelin and Richard Fellow, 'The State of Global Jihad Online', *New America Foundation*, 4 February 2013, http://www.newamerica.net/publications/policy/the_state_of_global_jihad_online#_edn.
12. Seventy-eight per cent of ex-burglars interviewed said that they strongly believed that social media platforms are being used by thieves when targeting property. See 'What's your Status?', *Friedland*, 26 September 2011, http://www.friedland.co.uk/EN-GB/NEWS/Pages/Whats-your-status.aspx.
13. Jamie Bartlett and Crl Miller, *@metpoliceuk* (London: Demos, 2013).
14. Sarah Clark, 'Police use Facebook to Identify Weapon Carriers', *The Journal* no. 18, 11 February 2009, http://www.journal-online.co.uk/article/5410-police-use-facebook-to-identify-weapon-carriers.
15. Richard Frank, Connie Cheng and Vito Pun, 'Social Media Sites: New Fora for Criminal, Communication, and Investigation Opportunities', *Public Safety Canada*, August 2011, http://publications.gc.ca/collections/collection_2012/sp-ps/PS14-5-2011-eng.pdf.
16. Mike Flacy, 'NYPD Creates Unit to Track Criminals on Social Networks', *Digital Trends*, 10 August 2011, http://www.digitaltrends.com/social-media/nypd-create-unit-to-track-criminals-on-social-networks/.
17. Xuning Tang and Christopher C. Yang, 'TUT: A Statistical Model for Detecting Trends, Topics and User Interests in Social Media', International Conference on Information and Knowledge Management 2012 (ACM: New York, 2012), pp.972–981.
18. Robert Wilson, Samuel Gosling and Lindsay Graham, 'A Review of Facebook Research in the Social Sciences', *Perspectives on Psychological Science* (2012), Vol. 7, No. 3, pp.203–220.
19. Hila Becker, Mor Naaman and Luis Gravano, 'Learning Similarity Metrics for Event Identification in Social Media', WSDM '10 *Proceedings of the Third ACM International Conference on Web Search and Data Mining*, pp.291–300.

20. Viktor Mayer-Schonberger and Kenneth Cukier, *Big Data* (London: John Murray, 2013).
21. Thereby avoiding a number of measurement biases often present during direct solicitation of social information, including memory bias, questioner bias and social acceptability bias. Social media is, by contrast, often a completely unmediated spectacle.
22. Examples include Deloitte's Sentscheck, and the Radian6 Dashboard
23. This is a focal research emphasis for Demos: 'Towards a Social Media Science: Tools and Methodologies', National Centre for Research Methods website, http://www.ncrm.ac.uk/research/MIP/socialmediascience.php.
24. John Bollen et al., 'Twitter Mood Predicts the Stock Market', *Journal of Computational Science* (2011), Vol. 2, No. 1, pp.1–8, http://www.relevantdata.com/pdfs/IUStudy.pdf.
25. Micol Marchetti-Bowick and Nathanael Chambers, 'Learning for Microblogs with Distant Supervision: Political Forecasting with Twitter', *Proceedings of the 13th Conference of the European Chapter of the Association for Computational Linguistics*, April 2012, pp.603–612, http://aclweb.org/anthology-new/E/E12/E12-1062.pdf.
26. Stuart Shulman 'Keeping Humans in the Machine Learning Loop' paper presented to the Social Text Workshop (closed), University of Birmingham, 28 March 2013.
27. Malcolm Sparrow, 'Network Vulnerabilities and Strategic Intelligence in Law Enforcement', *International Journal of Intelligence and Counter Intelligence* (1991), Vol. 5, No. 3, pp.255–274.
28. Rongjing Xiang, Jennifer Neville and Monica Rogati, 'Modeling Relationship Strength in Online Social Networks', *Proceedings of the 19th International Conference on World Wide Web*, 26–30 April 2010, pp.981–990, http://delivery.acm.org/10.1145/1780000/1772790/p981-xiang.pdf?ip=137.73.98.151&id=1772790&acc=ACTIVE%20SERVICE&key=C2716FEBFA981EF18894975F719C0CD09E871DDC3BD3EACA&CFID=357898972&CFTOKEN=77529903&__acm__=1378223380_e2b6dc5b3fbb965b38352dc518ad6133.
29. For example, Jianshu Weng, Ee-Peng Lim, Jing Jiang and Qi He, 'TwitterRank: Finding Topic-Sensitive Influential Twitterers', *paper presented at the Third ACM International Conference on Web Search and Data Mining*, 3–6 February 2010, http://ink.library.smu.edu.sg/cgi/viewcontent.cgi?article=1503&context=sis_research.
30. Rob Cross, Andrew Parker and Stephen P. Borgatti, 'A Bird's-Eye View: Using Social Network Analysis to Improve Knowledge Creation and Sharing', IBM Institute for Business Value, 2002, http://www.analytictech.com/borgatti/papers/cross,%20parker%20and%20borgatti%20-%20A_birds_eye_view.pdf.
31. Viktor Mayer-Schonberger and Kenneth Cukier, *Big Data* (London: John Murray, 2013).
32. Robert Nisbet, 'Review of Predictive Analytics: The Power to Predict who will Click, Buy, Lie or Die', *Smart Data Collective*, 27 March 2013, http://smartdatacollective.com/ericsiegel/113761/new-predictive-profession-odd-yet-newly-legitimate?utm_source=feedburner&utm_medium=feed&utm_campaign=Smart+Data+Collective+%28all+posts%29.
33. Hal Varian, 'Predicting the Present', *Google Insights,* http://ssl.gstatic.com/think/docs/predicting-the-present_articles.pdf; Xiaofeng Wang, Matthew S. Gerber and Donald E. Brown, 'Automatic Crime Prediction Using Events Extracted from Twitter Posts', *Social Computing, Behavioral – Cultural Modeling and Prediction, Lecture Notes in Computer Science* (2012), Vol. 7227, pp.231–238.
34. Nisbet, *Review of Predictive Analytics*.

35. See Jemima Kiss, 'Twitter Election Predictions are more Accurate than YouGov', *The Guardian*, 13 May 2010, http://www.guardian.co.uk/media/pda/2010/may/13/twitter-tweetminster-election for an overview of Tweetminster's study attempting to predict the UK 2010 general election.
36. Andreas Jungherr, Pascal Jürgens and Harald Schoen, 'Why the Pirate Party Won the German Election of 2009 or The Trouble With Predictions: A Response to Tumasjan, A., Sprenger, T. O., Sander, P. G., & Welpe, I. M. "Predicting Elections With Twitter: What 140 Characters Reveal About Political Sentiment"', *Social Science Computer Review*, 25 April 2011, http://ssc.sagepub.com/content/early/2011/04/05/0894439311404119.
37. Adam Sadilek, Henry Kautz and Vincent Silenzio, 'Modelling Spread of Disease from Social Interactions', *paper presented at the Sixth International AAAI Conference on Weblogs and Social Media*, 4–7 June 2012, http://www.cs.rochester.edu/~sadilek/publications/Sadilek-Kautz-Silenzio_Modeling-Spread-of-Disease-from-Social-Interactions_ICWSM-12.pdf.
38. See, for example, Heather Kelly, 'Police Embracing Tech that Predicts Crimes', *CNN*, 9 July 2012, http://www.cnn.com/2012/07/09/tech/innovation/police-tech.
39. 'Social Media and Prediction: Crime Sensing, Data Integration and Statistical Modeling', National Centre for Research Methods, http://www.ncrm.ac.uk/research/MIP/prediction.php.
40. Oliver Mason, paper presented to the Social Text Workshop (closed), University of Birmingham, 28 March 2013.
41. Nate Silver, *The Signal and The Noise* (London: Penguin, 2012), p.452.
42. Micol Marchetti-Bowick and Nathanael Chambers, 'Learning for Microblogs with Distant Supervision: Political Forecasting with Twitter', *Proceedings of the 13th Conference of the European Chapter of the Association for Computational Linguistics*, 23–27 April 2012, pp.603–612, http://aclweb.org/anthology/E/E12/E12-1062.pdf; Brendan O'Connor, Ramnath Balasubramanyay, Bryan R. Routledge and Noah A. Smith, 'From Tweets to Polls: Linking Text Sentiment to Public Opinion Time Series', *Proceedings of the AAAI Conference on Weblogs and Social Media*, May 2010, pp.122–129, http://www.cs.cmu.edu/nasmith/papers/oconnor+balasubramanyan+routledge+smith.icwsm10.pdf.
43. Vicki G. Morwitz and Gavan J. Fitzsimons, 'The Mere-Measurement Effect: Why Does Measuring Intentions Change Actual Behavior?', *Journal of Consumer Psychology* (2004), Vol. 14, No. 1, pp.64–74.
44. Stuart Sutherland, *Irrationality* (London: Pinter & Martin, 2007).
45. Michal Austin, 'Self-Deception and Social Media', *Psychology Today*, 6 May 2013, http://www.psychologytoday.com/blog/ethics-everyone/201305/self-deception-and-social-media.
46. Kit Eaton, 'There are More "Fake" People on Facebook than Real Ones on Instagram', *Fast Company*, 2 August 2012, http://www.fastcompany.com/1844504/there-are-more-fake-people-facebook-real-ones-instagram.
47. Jeffrey T. Hancock, 'Digital Deception: Why, When and How People Lie Online', in Adam Joinson, Katelyn McKenna, Tom Postmes and Ulf-Dietrich Reips (eds.), *Oxford Handbook of Internet Psychology* (Oxford: OUP, 2007), pp.289–302.
48. Eugene Santor Jr and Gregory Johnson Jr, 'Towards Detecting Deception in Intelligence Systems', Occasional Paper, University of Connecticut, http://www.dartmouth.edu/~humanterrain/papers/article.pdf.

49. See, for example, General Dutch Intelligence and Security Service, 'Jihadism on the Web', 2012, p.3, http://wwww.aivd.nl/@2872/jihadism-web; and J.M. Berger and Bill Strathearn, 'Who Matters Online: Measuring Influence, Evaluating Content, and Countering Violent Extremism in Online Social Networks', report published by the International Centre for the Study of Radicalisation and Political Violence (London: King's College London, 2013), http://icsr.info/2013/03/who-matters-online-measuring-influence-evaluating-content-and-countering-violent-extremism-in-online-social-networks/.
50. For instance, in deprived areas of Berlin, civil servants have increasingly used portable devices connected to database records when visiting care homes for the elderly and hospitals. These devices give constant, mobile access to databases, enabling public servants to understand the needs of individuals and families, track their previous contact, and check for problems and underlying issues that may have been recorded by other agencies. See Jeremy Millard, 'eGovernance and eParticipation: Lessons from Europe in Promoting Inclusion and Empowerment', *pa*per presented to the UN Division for Public Administration and Development Management (DPADM) Workshop: E-Participation and E-Government: Understanding the Present and Creating the Future, 27–28 July 2006, unpan1.un.org/intradoc/groups/public/documents/UN/UNPAN023685.pdf.
51. HM Government, *A Strong Britain in an Age of Uncertainty: The National Security Strategy*, (Government Stationary Office, October 2010), http://www.direct.gov.uk/prod_consum_dg/groups/dg_digitalassets/@dg/@en/documents/digitalasset/dg_191639.pdf.
52. This argument draws on the work of Susan Brenner: Susan Brenner, 'Gigatribe, Privacy and Consent', CYB3RCRIM3 Blog, 3 June 2011, http://cyb3rcrim3.blogspot.co.uk/2011/06/gigatribe-privacy-and-consent.html.
53. See HM Government, *A Strong Britain in an Age of Uncertainty*.
54. See Omand, *Securing the State*.

3
The Impact of Open Source Intelligence on Cybersecurity

Alastair Paterson and James Chappell

The way that organisations use information technology (IT) has changed significantly in the past two decades. The emergence of new trends, such as social media, cloud computing (in its various guises) and the consumerisation of IT, have all had profound impacts on the way in which organisations seeking greater efficiency, cost savings and new business opportunities share, manipulate and exploit information. Critically, from a cybersecurity perspective, this nexus of forces has caused organisations and their employees to expose more information online than ever before.[1] Although much of this information is benign, some of it may constitute sensitive data which can then be exploited by increasingly sophisticated cybercriminals who use it to support their efforts to penetrate organisations' IT systems and steal valuable data or disrupt services. This chapter starts by summarising the recent merging of different forms of IT and how this has left organisations susceptible to OSINT-supported cyber attacks. It then goes on to dissect these new threats, outlining organisational strategies for risk mitigation, before discussing emerging challenges.

The blurring of boundaries

Three major developments in ICT are currently shaping enterprises across all sectors – social media, cloud services and mobile devices. The widespread adoption of these new technologies has revolutionised how individuals and organisations access, store and share information.

Social media – the power and reach of which is discussed in Chapter 2 – is the most recent iteration of user-generated content and provides for unprecedented levels of expression and interaction. The recent explosion in social media usage, for example, has propelled Facebook, the world's largest social network, to 1.11 billion monthly active users as of March 2013, of which 751 million connect via mobile devices.[2] Similarly, LinkedIn, a social networking website for professionals, had expanded to over 200 million users as of January 2013, while Twitter reached the same milestone in December 2012.³ Recently launched Google+, a social networking and identity service that competes with Facebook, had 359 million active users on 1 May 2013.⁴ Crucially, while these major social networks have all been developed and launched in the US, their geographical spread has expanded almost as quickly as their membership bases: for example, approximately 79 per cent of Facebook users are now from outside North America.

In addition to these truly global social networks there are a significant number of services with a national or regional focus as innovative developers attempt to respond to a perceived gap in the social media market. A good example here is China, where the prohibition of US-based services, such as Facebook, LinkedIn and Twitter, has led to the establishment of a range of local replacements, such as QZone, Tencent Weibo and Sina Weibo, with each now boasting hundreds of millions of users. Of course, socially generated content encompasses more than that shared via social networks; as of October 2013, WordPress, one of the most used blogging platforms, was hosting over 71 million blogs globally.⁵

The wide adoption of social media is linked to the myriad of social, economic and political benefits that it provides to organisations and individuals in terms of the potential to make new and potentially lucrative connections. From political protest to cultural awareness, social media can facilitate the exchange of ideas and thoughts. In the business world, this function is a valuable one. A growing number of companies now promote themselves on social media sites, using them as a means of connecting with existing and new customers and partners. There is a growing realisation that the exploitation of social media can provide a competitive advantage, so businesses are actively seeking to grow and enhance their digital footprint, with more and more corporate discussion occurring online.

A second major development comes in the form of cloud computing. The term 'cloud services' is commonly used to describe a recent trend towards large-scale distributed computing, where multiple companies are tenants of a single infrastructure and benefit from a common set of IT services from a single provider. Cloud services have proved popular with private individuals, with commercial providers such as Dropbox offering convenient cloud-file hosting. The real market here, however, is the commercial one. Analysts predict that the public cloud-services market will grow by some 18.5 per cent in 2013 to total USD131 billion worldwide.⁶ For many organisations the adoption of cloud services has direct financial benefits: storing files in a virtual repository rather than a physical one can significantly reduce costs. Another important driver is ease of access to information. No longer constrained by the location of files, employees can access company files and information from anywhere in the world. Cloud services have thus provided companies with real incentives to grow their virtual presence, and increasing amounts of corporate information, once held on servers within the company, is now held off premises.

Of course, embracing the 'cloud' brings its own challenges. From a security perspective, for example, it is now only necessary for cybercriminals to penetrate the security system of one network to gain access to multiple companies' sensitive proprietary information. While few in number, such breaches have already occurred. In July 2012, for example, usernames and passwords stolen from other websites were used to sign in to a number of Dropbox accounts.[7] In particular, Dropbox users began to receive spam email when one of the stolen passwords was used to access an employee Dropbox account containing a project document with user email addresses. The situation was quickly resolved but it illustrates the security risks posed by the increasing reliance on cloud services.

Moreover, the ubiquity of connectivity has blurred the boundaries between work and personal life. There has been an explosion in the use of smart phones and tablets for work purposes, and many organisations have embraced this development, taking steps to provide appropriate security measures for personal devices. Indeed, in some cases, organisations are embracing the mobility offered by portable devices by rolling out their own programmes of tablets. For example, in November 2012 it was reported that Barclays ordered approximately 8,500 iPads for front-office staff in its 1,600 branches, becoming one of the first UK banks to do so.[8] Here again there are clear benefits to employers in empowering their workforce with these devices. Although, as with cloud services, the consumerisation of IT has engendered a fundamental shift in the way that corporate data are handled. This reflects a broader shift from a tightly controlled model to a widely distributed one in which personal devices either hold or provide access to potentially sensitive corporate information. Clearly, this can pose a significant problem in cases where employees are accessing corporate information on personal devices without organisational sanction, or where security measures applied by companies are inadequate.

Cyber attacks: A growing business

The information revolution of the late twentieth century has been driven by a march of significant improvements in our capacity to store, compute, exchange and receive information. This has, in turn, created a wealth of opportunities for commercial stakeholders, a few of which were mentioned above. At the same time, however, these changes in the information landscape have presented malicious actors with increasing incentives and opportunities to target commercially valuable online and digital information, which is often poorly secured. This is reflected in security industry reports which detail large increases in cyber attacks. For example, a 2013 report by Symantec, a company specialising in information security online, revealed a growth in targeted cyber attacks of 42 per cent in the previous 12 months.[9] Moreover, commercial research suggests that the economic costs of cybercrime are significant, and, even if the exact figures are disputed, most commentators agree that this is a big problem that is getting bigger. In a 2011 report published by BAE Systems Detica in conjunction with the UK Cabinet Office, the annual cost of cybercrime to UK business was estimated to be in the region of GBP21 billion, although these figures have been hotly disputed.[10] This estimate came after the UK National Security Strategy (NSS) and Strategic Defence and Security Review (SDSR) released in October 2010 'promoted cyber security to a Tier One risk to national security'.[11] The high status of the risk was 'reinforced by the UK government's allocation of £650 million to cyber security and resilience'.[12]

Furthermore, it has become easier for criminals to launch cyber attacks. The traditional barrier of deep technical understanding has been eroded by the emergence of easily deployed 'point-and-click' tools. In 2010, for example, *Forbes* revealed details of an innovative approach to Distributed Denial of Service (DDOS) attacks.[13] A particular website – IMDDOS.org – offered paying customers the opportunity to download an application that gave them temporary use of a collection of hijacked computers. Essentially, this process allowed the customer 'to target their own DDOS attacks with just a few keystrokes'.[14] This phenomenon has been accompanied by a rise in 'hacking as a service', experienced hackers offering their services for a fee via both by invitation-only and on open fora.[15]

The drivers and rationale for the attacks have also evolved. Theft of intellectual property, state-level espionage and large-scale organised crime are commonplace today whereas, historically, many of the breaches and much of the disruption were committed as simple fraud, or sometimes just for entertainment purposes. The period from the late 1990s to the mid-2000s, for example, was one dominated by opportunistic crimes targeting mass audiences, attempting to infect computers or coerce users into giving away vital information or conducting a fraud. 'Phishing' emails and generic computer viruses were common.[16] However, as users become more aware of these methods of fraud, their effectiveness is reduced.

In recent years, cyber attacks have thus become far more targeted, with Cisco, a network security and infrastructure vendor, reporting in June 2011 a 300 per cent rise in 'highly personalised' attacks.[17] The shift here is driven, to a large extent, by economics. There has been a growing realisation among cybercriminals that the opportunity cost of spamming may not be worth the rate of return due to increases in both anti-spam software and user awareness. Instead, more time and effort are being devoted to 'different types of targeted attacks, often with the goal of gaining access to more lucrative corporate and personal bank accounts and valuable intellectual property'.[18] With these targeted attacks holding the promise of much greater rewards, attackers are now investing considerable time and effort in hostile reconnaissance in order to both select appropriate targets and maximise the effectiveness of their attacks.

Attacks may now only target a handful of individuals, handpicked because of the access they may have to information desired by the attacker. The computers and systems used by the targeted individuals are subject to reconnaissance by the attacker to establish what protection is in place, with a view to constructing a tailored package of malicious code that would evade the defences of that network and computer. This much achieved, the attacker then conducts further reconnaissance to establish a more intrusive and potentially long-term presence on the network. In general terms, the quantity and effectiveness of these targeted cyber attacks have advanced, and targeted attacks are now a cause for major concern. In June 2012, for example, the UK Security Service claimed that it is now encountering 'astonishing' levels of targeted cyber attacks on UK industry.[19]

OSINT and hostile reconnaissance

As mentioned, in their efforts to mount a targeted cyber attack, attackers are placing increasing emphasis on the reconnaissance process; the Council of E-Commerce Consultants estimates that 90 per cent of the time involved in a targeted cyber attack is now spent in the profiling phase.[20] Crucially, attackers are increasingly looking to exploit open source information relating to an organisation to support their attack.

This information may be inadequately secured in the cloud, or compromised by the lax security measures around mobile technology platforms. Crucially, however, information is often readily available in open sources. For example, significant quantities of data, especially that posted on social media sites, can be defined as open source – being publically accessible – and can, when paired with information intentionally put out by a company about the organisation, be linked back to specific companies through its employees, clients, partners and suppliers, providing an often comprehensive insight into the company. The availability of this information is a direct consequence of the aforementioned rise of social media and cloud computing, and was not readily accessible online before the advent of these platforms. If all of the information available online about an individual or organisation is thought of in terms of a digital footprint, then sensitive and unintentionally exposed information that can potentially be used to support a cyber attack may be regarded as a 'digital shadow'. It is this that criminals seek to identify, map and, ultimately, exploit for profit.

Tools and techniques for gathering sensitive information

The basic approach to conducting a cyber attack has remained unchanged since the early days of computer network security. The first step in the attack process is to conduct research regarding the target, and to ascertain and probe the IT assets of the target organisation. The goal here is to identify points of weakness in an organisation's IT architecture. Indeed, so well established is this phase of the cyber attack that professional network-penetration testing firms are often commissioned to conduct 'Red Team' exercises with a view to identifying system weaknesses that might be exploited by attackers. Cognisant of the amount of information that is exposed online, often unwittingly, both the hacking and the security communities have developed increasingly sophisticated tools to help to identify and exploit, or mitigate the risks associated with, this wealth of data.

A major challenge in this regard relates to the fact that advancements in IT system security form part of a circular process: the tools and techniques developed by security specialists for the purpose of defending computer networks by helping network administrators to learn about potential security weaknesses may also be used by attackers for malicious purposes.

These tools are often command-line-driven utilities that are designed to explore network interfaces. For example, free tools such as ZMAP and NMAP can be used to gather information about a network's externally facing interfaces, the points of entry for an attacker. The researchers behind ZMAP, for example, a modular and open source network scanner specifically designed to perform comprehensive Internet-wide research scans, claim that 'recent studies have demonstrated that Internet-wide scanning can help reveal new kinds of vulnerabilities, monitor deployment of mitigations and shed light on previously opaque distributed ecosystems'.[21] This is a highly invasive technique and as such security professionals will seek the network owner's permission before employing such tools. The Internet's own domain name system (DNS) mapping system also provides opportunities to discover potential entry points for an attacker onto a target network, with tools created to both guess the addresses of a company's systems and explore known name systems. These openings can be exploited if the devices at the addresses identified are not configured correctly, leaving open ports, or are not 'patched' with the latest software versions that eliminate known vulnerabilities. Of course, there are websites that attempt to capitalise on this process. For example, www.shodanhq.com has created a search engine of exposed computers on the Internet.

Although developed by security professionals, footprinting and enumeration tools are more and more frequently employed by attackers to map the external technology infrastructure of a target organisation. Another relevant example is the recent emergence of self-styled 'open source intelligence toolkits', such as Maltego, a product of South African company Paterva.[22] Maltego attempts to provide users with an 'aggregation of information posted all over the internet – whether it's the current configuration of a router poised on the edge of your network or the current whereabouts of your Vice President on his international visits'. Clearly, however, the value of this toolkit to companies is matched by its value to attackers conducting hostile reconnaissance. From a simple item of 'seed' data, much can be learned about the target, its make-up and its weak points.

At the beginning of its development, the Maltego toolset followed the traditional route of computer network enumeration in a manner similar to the command-line tools mentioned above. However, this has quickly spread to encompass a range of sources and to act in a manner more akin to professional intelligence network analysis tools, such as IBM's i2 Analyst's Notebook, or the intelligence toolsets provided by Palantir.[23]

The computer security community has long realised the power of conventional search engines in identifying security flaws in IT systems. In *Google Hacking for Penetration Testers*, Johnny Long outlines how the advanced features of Google's search engine – available to all – can help to locate computers and targets of interest to an attacker.[24] For example, the simple search 'inurl:/cgi-mod/index.cgi' returns a list of web servers that employ old Common Gateway Interface protocols, which is an indication of poor maintenance and patching, but can also indicate sites which have significant weaknesses.

Gathering information online is no longer a matter of time-consuming, individual searches. Rather, more sophisticated approaches have been developed that run searches across multiple search engines to bring together lists of servers and infrastructure that might be vulnerable. For example, the Google Diggity project, established in 2010, looks to evolve the art of 'Google hacking' through the creation of Windows desktop tools that carry out many thousands of automated searches across major search engines.[25]

Data that are stored in computer files are commonly enhanced with additional data about where and how the file was processed. These data are referred to as 'metadata', or data about data. Metadata are commonly used by security researchers in order to discover weaknesses in network or system security. For example, the files stored on a webserver can include details about internal file structures, usernames and other internal company resources. More specifically, in the case of an image file, a great deal of metadata can be included, such as the location where a picture was taken, the capturing device, and the date and time. Metadata are yet another revealing source of information potentially of use to hackers in planning a cyberattack.

Other online information sources that may be used for hostile reconnaissance include network information providers and social networks. Network information providers maintain records of internet protocol (IP) addresses registered to different companies, allowing potential attackers to ascertain which computers belong to which organisations. Furthermore, these are easily accessible resources as they underpin the Internet itself. They offer online search tools that can enable users to identify users of websites easily, and without technical expertise. Social networks also provide a source of easily accessible information that, when aggregated, can paint a rich picture of an organisation's IT setup. In this case, attackers may search for IT department and staff profiles to gain information about the structure of the department and areas of staff expertise which, in turn, may provide an insight into the software being utilised within the organisation.

In addition, there are now a variety of information aggregators, which trawl social media sites for information. Examples here include www.spokeo.com, www.123people.com and www.192.com. As well as picking up organisational IT-specific information, they will also collection personal data relating to individuals, such as home addresses, birth dates and educational history. This can be useful for attackers seeking to guess online passwords and associated reset questions, or in socially engineered attacks, which will be discussed below.

Ultimately, the range of sources that can assist attackers in planning a cyber attack is large and diverse. And, inevitably, as we spend more and more of our time online, the amount of information populating these sources grows accordingly. Given this wealth of sources, it is not surprising that attackers are investing significant resources and time to pick data sources that serve their targets' interests. For example, in the authors' own experience during a recent project involving an established mining company, software support forums for a well-known and frequently used geological service were found to contain a number of spoof accounts. It was later shown that these accounts were associated with a group seeking to target the internal IT systems of the mining firm and the participants of the forum became the targets of a socially engineered online attack.

The above example is indicative of a growing trend. Social networks have provided new opportunities for attackers to gain access to sensitive information. By using this information and by impersonating individuals and organisations, attackers can more easily gain access to a target. After identifying a target, attackers can either create a spoof account of an associate of the individual, or gain access to an existing account by compromising the security of the social media site. Once the attacker has access to the account, they are free to send messages requesting sensitive information, or to post corrupted hyperlinks that, when clicked, will compromise the security of target computers, giving the attacker access to the desired network. The strategy of impersonating a known and trusted associate means that the chances of targets clicking on corrupted links or revealing sensitive information are dramatically increased.

Categorising targeted cyber attacks

The tools and techniques employed in hostile reconnaissance feed into a range of targeted cyber attacks that adversaries can launch against both individuals and organisations. These attacks vary in terms of their route, sophistication and end goal. This section will consider a number of the most prominent attack types currently in use, including spear-phishing, tailored malware injection and facilitated physical attacks.

Spear-phishing

Spear-phishing involves the targeted spoofing of seemingly legitimate emails or online messages in an attempt to trick a key individual within an organisation, such as a chief executive or network manager, into carrying out an action that compromises the security or integrity of a network. This differs from 'traditional' mass spamming due to the level of reconnaissance and customisation that the attack involves. Almost indistinguishable from a legitimate message, spear-phishing attacks are able to circumvent standard network security defences, such as email filters, which are set up to block generic spam and fool the target into treating the message as genuine. By opening an attachment or clicking on a link, the target then inadvertently opens the way for malicious software to be installed on their computer. Spear-phishing style attacks are commonly used by cybercriminals for the purposes of financial theft, with studies showing that if implemented correctly they can net up to ten times the profits of mass spamming.[26]

Another related, commonly used technique is 'water-holing'. With this approach an attacker will invest considerable resources in researching the browsing habits of a target. Having gained an understanding of the online interests and favoured websites of one or more key employees, the attacker will then seek to breach a relevant third-party website and install malware that is activated when the employee visits the page. If this occurs when the employee is using a company computer, the infection spreads to the broader corporate network of the target. As with social media impersonation, gaining a detailed understanding of the target in advance of the attack can greatly increase the attackers' chances of successfully compromising the network. Moreover, these techniques are often used simultaneously in order to increase the attacker's chances of penetrating the target network.

Tailored malware injection

In addition to seeking out sensitive information about individuals' and company processes, attackers may also gather data on a company's IT systems and software in order to tailor the malware used in an attack. This can provide insights into both the likely vulnerabilities in that network and the likely defences that the attacker will meet when trying to gain access to the system. A common approach in conducting this type of attack is for the attacker to research the business profiles of IT staff that will share information about skills that they have acquired to deliver the technology in use in an organisation. Having understood the layout of an organisation's network and software versions, the attacker is able to inject malware with increased chances of success in bypassing security measures. An example of this approach is detailed in a blog post by security company RSA, published after a cyber attack on the US security firm in 2011.[27]

'Doxing' and facilitated physical attacks

In some cases, cyber attacks can cross over into the real world in a physical and truly frightening way. In an effort to target their cyber attacks, perpetrators will employ hostile reconnaissance to gain a comprehensive understanding of the target's digital shadow. The focus here can be professional, personal or both. In a professional context, relevant information making up a digital shadow might include key employee information, usernames, passwords, software versions, confidential documents, and network and security settings. Moreover, many organisations are unaware that the aforementioned information is sensitive: there are, for example, websites that actively encourage network administrators to publish their employers' network diagrams.[28] In a personal context, a digital shadow includes information about the target's acquaintances, family members and habits, and likely location.

Whatever the focus, once the information is gathered, attackers can choose to exploit it directly or release it to the wider community in so-called 'doxing' attacks. These are typically targeted at specific individuals in an organisation who hold power, have access to resources or are likely to respond to coercion attempts through activities such as kidnap or ransom. Sensitive information about the target is published online to encourage others to do harm to those individuals. A good example of this is Operation Wall Street, launched in early 2013 by the loosely organised collective of cyberactivists known as Anonymous. The stated aim of the attackers was to capture and publish online personal information about the chief executives of major US banks.[29] Of course, these attacks can potentially lead on to more serious criminal acts, potentially facilitating physical attacks or kidnapping in extreme circumstances.

Given the range of resources available to cybercriminals and the variety of attacks at their disposal, how can the threat posed by cyber attacks be reduced or mitigated? What measures can individuals and organisations take to secure their information and their networks?

Mitigating the risks from OSINT-directed cyber attacks

For many years the network perimeter has been the vanguard of electronic defence. The approach adopted by commercial organisations has been to build the walls high and defend the boundary. Security measures such as anti-virus software, firewalls and intrusion-detection and prevention systems have all played a significant role in protecting commercial data. However, as mentioned above, the consumerisation of IT and, in particular, the growth and widespread adoption of social media and cloud services have undermined and eroded the traditional barriers separating sensitive information from the public domain. More and more information is now publicly accessible through the open Internet. Here a historical parallel can perhaps be drawn with the gradual dismantling of city walls or, in some parts of the world, state boundaries over time in order to enhance the free flow of goods and information.

This erosion of perimeters, the breaking-down of boundaries between corporate networks and the public sphere, is further accentuated by the growth and increased complexity of organisational supply chains: 'companies establish partnerships which require interconnected systems and networks to participate in electronic commerce and to allow them to operate within each other's supply chains'.[30] Patterns of globalisation and trends towards outsourcing have lent additional momentum to the process. In this new environment, traditional defensive methods are unable to withstand the new, more sophisticated and targeted cyber attacks that are facilitated and driven by readily accessible open source information. Consequently, it is becoming increasingly necessary for companies to rethink their IT security policies. Some companies, for example, have responded to these security concerns by placing constraints on the use of the latest online technologies and resources. Others have chosen to prohibit the use of social media and cloud services in the workplace.

However, these repressive approaches are not all-encompassing. Company employees, for example, continue to access social media from their homes and from their personal mobile devices. Furthermore, access to social media, and the publicity and market research that it can provide is now a strategic requirement for forward-facing businesses and individuals. In any case, preventing company employees from using social media does not mean that partners, clients and suppliers – all of whom may hold sensitive corporate data – will also do so.

As with many security problems, there is no single solution that will prevent open source information about organisations from being gathered and used in a targeted cyberattack. Crucially, however, there are a number of steps that companies can take to mitigate the risks posed by such an attack, including the development of a comprehensive and security-oriented social media policy and associated training, and monitoring of potential vulnerability.

Social media policies are now becoming commonplace for public- and private-sector organisations alike. These are increasingly being introduced as part of the corporate governance and compliance processes in an organisation, and seek to clarify the dos and don'ts for employees who wish to engage with social media, both on behalf of the organisation and in a personal capacity.[31] As part of the policy process, employers tend to consider how an organisation represents itself online, as well as which elements of the organisation have the authority to disseminate information publically. Crucially, from a public relations point of view, these policies often seek to determine some sort of acceptable posting and interaction rules for employees, partners and their associates. In this respect, social media policy is simply an extension of the existing 'acceptable use' policies that govern IT practices in most businesses today. These policies provide a legal and regulatory means to enforce lawful use of an organisation's IT systems and ensure that there is a referenced set of rules and associated guidance to which employees can be directed.

However, in order for a social media policy to have a direct impact, clear communication and delivery are critical. The publication of a long and complex document, full of technical jargon, is insufficient to meet employees' needs. In this context, organisations have adopted a range of training approaches to ensure that social media policies are understood and widely adopted, and contribute to the development of a broader culture of compliance. A good example of an awareness campaign is provided by the UK Ministry of Defence (MOD). The MOD used a series of short videos to highlight the importance of responsible and appropriate engagement with social media. Aimed at service personnel and their families, the initiative was dubbed 'Tweets Cost Lives' – a play on the slogan that adorned informational posters during World War II: 'Careless Talk Cost Lives'.[32] It reflected growing concerns within the MOD of the risks posed by careless use of social media. For example, careless online discussion could potentially give away sensitive details regarding the position of strategic military assets.

Another important means of offsetting the risk posed by cyber attacks – and particularly the effectiveness and value of hostile open source reconnaissance being carried out on a company – is for the organisation to adopt the perspective of a potential attacker. By considering the company from an attacker's perspective, those working in the company can identify, highlight and remove information that might pose a risk or threat. This approach can be enhanced through the use of professional penetration-testing firms, many of which now offer desk-based audits or red teaming, where a team of technically minded specialists attempt to attack the client's organisation. However, this process can demand considerable time and resources: identifying the most subtle vulnerabilities may require a highly sophisticated attack.

By far the most effective method of mitigating the risks posed by cybercriminals is to adopt a holistic approach to the monitoring and analysis of the people, processes and technology involved in a business. There are now a range of Internet monitoring services designed to help to identify a variety of important vulnerabilities and red flags:

- weaknesses in the security of the organisation's network perimeter;
- compromised hosts or malicious code targeting an organisation;
- evidence of beaconing behaviour from networks;
- traffic analysis to identify poorly configured devices attached to the corporate network;
- analysis of data that have inadvertently leaked from the business;

- monitoring of storage and cloud services for accounts impersonating key individuals;
- publication of proprietary documents or copyrighted materials;
- evidence of attacks or groups organising to attack a business;
- activist group monitoring services.

Monitoring also provides assistance with internal compliance, as employees who are aware that monitoring is in place as a routine activity are less likely to 'bend the rules' or turn a blind eye to the policies in place.

More focused monitoring of key individuals can also be beneficial. Individuals who hold high-risk or highly public positions in a business, for example, can benefit from targeted audits of their online information. Monitoring and even managing the digital shadow of these key individuals can do much to reduce their potential vulnerability to targeted attacks. This can also offset the physical risks that information regarding location, associations, attitudes and other aspects of an individual's daily life can engender. It was for this reason that personal details relating to then incoming head of the UK Secret Intelligence Service (MI6), Sir John Sawers, were removed from Facebook in 2009.[33] Sir Sawers' wife had, through posts on Facebook, revealed information about the couple's children as well as the location of the family apartment. Exposures such as this can be avoided, or at least dealt with before the information becomes a problem, by constantly monitoring key individuals and their information.

Managing digital shadows

As technology and communications continue to advance and permeate our lives, more and more of our information finds its way online. In this context, the Internet serves as an enormous and enduring repository of information. The Internet is, by its very definition, a resilient infrastructure designed to keep information available, even when individual communications degrade, or become disrupted. This reality fuels the notion that it is impossible to remove information completely from the Internet. Certainly, it can be difficult to do so; paradoxically, attempts to hide, remove or censor information may well attract unwanted attention, provoking a cascade of replication as commentators and observers turn a spotlight on the information in question. This almost circular process is often termed the 'Streisand Effect', after the actress Barbara Streisand. She 'sued the California Coastal Records Project [CCRP], which maintains an online photographic archive of almost the entire California coastline, on the grounds that its pictures included shots of her cliffside Malibu mansion, and thus invaded her privacy'.[34] The lawsuit attracted enormous attention and links to the pictures spread across the Internet. Ironically, 'as the links proliferated, thousands of people saw the pictures of Ms Streisand's house – far more than would otherwise ever have bothered to browse through the CCRP's archives'.[35]

Of course, the celebrity status occupied by Streisand makes hers an exceptional case. The vast majority of information that might be used to support a cyberattack would not draw much attention at all. Indeed, much of it is seemingly benign information. The social profiles of employees, the publication of IP addresses and the inadvertent publication of company documents are not issues that tend to attract much interest beyond the small circles of interest (including cybercriminals). In these cases, then, it becomes much easier to manage digital shadows by removing or adjusting offending or potentially sensitive information.

One way to approach the problem is to track back and identify the original publisher of the material in question. In the case of internal systems, employees or partners, the publication of potentially sensitive information is often inadvertent or unintentional. A polite and timely request to the relevant IT department or the person responsible for publishing the information online is often enough in the majority of cases. In cases where there is no existing relationship with the publisher, there may be a cost associated with removing the information. In both cases, those publishing information online must adhere to data-protection laws, and this can work in favour of the individual requesting the withdrawal or modification of information.

In the unusual cases where a third-party publisher is hostile or refuses to abstain from publishing sensitive material, it is possible to neutralise the effect of the publication. Internet reputation firms or online review firms can publish confusing and misleading information within or near the original content, creating misinformation and making it difficult for an independent reader to discern the original or genuine piece of information. Parallels can be drawn here with the use of counterintelligence techniques in the intelligence world.

Lastly, the ability to find information plays an important role in maintaining its veracity. Most major search engines offer brand holders a toolkit or access to application commands to help to influence rankings in search engines. This discipline of manipulating search rankings is known as search engine optimisation (SEO) and is used predominately by marketing firms to increase the sales of a product or service by ensuring that it appears in the first page of a set of search-engine results. For those interested in minimising their digital shadow, however, the technique can be reversed and used to 'de-optimise' or 'de-emphasise' information and bury it in the depths of search engine results.

The future of digital shadows

The rate of change and progress in the world of information and cybersecurity means that it is almost impossible to predict precisely how our digital shadows and their associated risks will grow and evolve over the coming years. There are, however, a number of upcoming trends that are worthy of consideration, including issues relating to the freedom of the Internet, and the increasing use of the Internet in everyday life, such as through wearable technology and smart homes.

Legislative developments will undoubtedly have important impacts in this area. Notably, they are likely to impact on the size of both individual and corporate digital shadows. Legislation relating to the Internet has been struggling to keep up with technological developments and must balance the need for regulation against the freedom of the Internet and the people who use it.

This is a complex area that will differ along geographical lines; different laws will be passed in different jurisdictions. For example, in the European Union (EU) the much discussed right 'to be forgotten', part of a wide-ranging overhaul of the European Commission's 1995 Data Protection Directive, may require technology companies to delete all personal information that they hold on individuals if requested to do so by the subject of the information.[36] In practice, this may prove extremely complicated since many companies have a business model that relies heavily on the gathering and storing of customer information, and in many cases the information is passed on to third parties. It is not yet clear what form this legislation will take or when it will be enacted.

Furthermore, in the US, several legislative bills have recently been proposed that will impact the regulation of information on the Internet. In particular the Cyber Intelligence Sharing and Protection Act (CISPA) (introduced in 2011 and, at the time of writing, stalled in the Senate) and the Stop Online Piracy Act (SOPA) (introduced in 2011 and currently under consideration in the House of Representatives) have been proposed to help to prevent online piracy and counterfeiting, and to help the sharing of cybersecurity intelligence. However, they have proved controversial and are seen by some as a means for the government to enforce tighter control on the Internet – controls with a bias towards industry.[37] It is not yet apparent what impact, if any, these or related acts will have on corporate digital footprints.

A second trend likely to impact on the nature of digital shadows is the increased use of wearable computing and life-logging technologies. This emerging market is currently thought to be worth somewhere in the region of USD3–5 billion, a figure that may grow by a factor of ten in the next two to three years.[38] At a basic level, devices such as the fitness and lifestyle trackers created by companies such as Nike, Fitbit and Jawbone all record information about our health that can then be shared on social networks. Other devices, such as the Zeo Headband, monitor and store information about our sleep patterns. Although these all seem like benign examples and can promote health and wellbeing among users, the next generation of wearable computing will likely go further and may well have the potential to record and publish corporate data in some form. For example, researchers are already experimenting with electronic fabrics that could allow our clothes to monitor, sense and record the environments around us.[39] Google Glass is probably the best-known recent advancement in wearable computing. Essentially, this is a computer worn as a pair of glasses, augmenting the user's vision with a small display that can show information in a hands-free format. Google Glass has many tools and applications, including the ability to record videos of everything that the wearer can see. Clearly, this has many potential security implications: in a context where Google Glass or other such products are always on, there is the potential to gather an almost limitless amount of valuable information – records of movements, meetings and social interactions, personal identification numbers (PINs), passwords and other sensitive material with which the user interacts. From, say, a marketing perspective, this would constitute a veritable goldmine of information. From a security perspective, however, this would be a repository of potentially sensitive and possibly vulnerable information.

Google Glass runs the Android operating system. Given that 2013 has seen targeted malware attacks on Android phones that steal phone numbers and text messages, it follows that cybercriminals will also seek to target these devices of the future.[40] Indeed, it is alleged that the security features of the beta version of Google Glass have already been breached.[41] Use of these devices also requires the users to trust companies such as Google to respect their privacy, something that might prove difficult given that the company has already experienced controversy relating to privacy, including being fined for illegally recording information from unsecured WiFi networks.[42]

The world of videogaming offers another relevant example of the security threats posed by technological advances. The next generation of videogame consoles includes the recently announced Microsoft XBox One. This promises to be a major technological advancement in many areas, but there have already been suggestions that features such as always-on listening and motion detection through video capture could be used for surveillance purposes.[43] As with Google Glass, a breach of the security of the XBox One could lead to very sensitive personal information being leaked onto the Internet, information that could be used in a targeted cyber attack.

By 2020 it is estimated that there will be some 50 billion devices connected to the Internet.[44] The benefits from the 'Internet of things' are often portrayed in terms of convenience for consumers: fridges that order milk before it runs out, houses that turn the heating and lights on just before you get home, among other things. Of course, this extends beyond the home into future smart cities, smart traffic, smart meters and a host of other developments touching almost every aspect of our lives. However, the proliferation of these 'smart objects' may simply serve to make us exponentially more vulnerable to cyber attacks.

Conclusion: Evolving and growing shadows

For all of their benefits, the rise of social media, cloud computing and the consumerisation of IT have caused organisations and their employees to expose more information online than ever before. Some of this information may expose critical information to increasingly sophisticated cybercriminals performing research on specific organisations to understand how best to break in and steal data or cause other forms of harm.

Organisations seeking to become more efficient and cost-effective while at the same time exploiting new business opportunities are keen to embrace and profit from technological advancement. However, in this new threat environment, businesses must also investigate new approaches to managing the risks that technological advancement brings. In this context, it is crucial that organisations educate their employees, develop comprehensive and dedicated policies and training, and monitor their digital footprints, for risks will thrive in this new world.

This chapter has provided an introduction to some of the opportunities and risks posed by recent technological developments, and social and business practices in the cyber domain. It has also sought to provide some examples of risk-mitigation techniques. As we look to the future, one thing is clear: our individual and collective digital shadows will continue to evolve and grow over time. We must simply make sure that they do not overwhelm us.

Notes

1. Chris Howard, 'Engage, Learn, Create and Disrupt With the Nexus of Forces', *Gartner*, October 2012, http://www.gartner.com/resources/239600/239602/engage_learn_create_and_disr_239602.pdf.
2. 'Facebook Newsroom – Key Facts', Facebook, http://newsroom.fb.com/Key-Facts.
3. Deep Nishar, '200 Million Members!', LinkedIn Blog, 9 January 2013, http://blog.linkedin.com/2013/01/09/linkedin-200-million/; Tony Wang, 'There Are Now More than 200M Monthly Active @twitter Users', Twitter, 18 December 2012, https://twitter.com/TonyW/status/281056150651293697.
4. Thomas Watkins, 'Suddenly, Google Plus Is Outpacing Twitter to Become the World's Second Largest Social Network', *Business Insider*, 1 May 2013, http://www.businessinsider.com/google-plus-is-outpacing-twitter-2013-5.
5. John Rampton, 'Blogging Stats 2012 (Infographic)', Blogging.org, 13 July 2012, http://blogging.org/blog/blogging-stats-2012-infographic/; 'Wordpress Sites in the World', WordPress, http://en.wordpress.com/stats/; Alex Konrad, 'Salesforce's Listening Tools Now Help Brands Track China and Russia's One Billion Social Users', *Forbes*, 7 November 2013, http://www.forbes.com/sites/alexkonrad/2013/07/11/salesforce-radian6-china-russia/.
6. 'Gartner Says Worldwide Public Cloud Services Market to Total $131 Billion', *Gartner Press Release*, 28 February 2013, http://www.gartner.com/newsroom/id/2352816.
7. 'Security Update and New Features', Dropbox Blog, 31 July 2012, https://blog.dropbox.com/page/4/?page=%2Fsecurity-update-new-features%2F.

8. Charles Arthur, 'Barclays Ipad Order Could Open Door to Wider Bank Use', *The Guardian*, 26 November 2012, http://www.guardian.co.uk/technology/2012/nov/26/barclays-ipad-order-bank-use.
9. Highlights from 2013 Internet Security Threat Report from '2013 Internet Security Threat Report, Volume 18', Symantec, http://www.symantec.com/security_response/publications/threatreport.jsp.
10. BAE Systems Detica, 'Cyber Risks Insurance: The Challenge and the Opportunity', *Security Horizons Briefing*, 24 November 2011, http://www.baesystemsdetica.com.au/getattachment/Research/Publications/Cyber-risks-insurance-the-challenge-and-the-oppor/Cyber_Risks_Insurancefinal_pdf.pdf.aspx.
11. Paul Cornish, David Livingstone, Dave Clemente and Claire Yorke, 'Cyber Security and the UK's Critical National Infrastructure', *Chatham House Report*, September 2011, http://www.chathamhouse.org/sites/default/files/public/Research/International%20Security/r0911cyber.pdf.
12. Ibid.
13. Andy Greenberg, 'Chinese Botnet Sells Point-And-Click Cyberattacks', *Forbes*, 13 September 2010, http://www.forbes.com/sites/andygreenberg/2010/09/13/chinese-botnet-sells-point-and-click-cyberattacks/.
14. Ibid.
15. 'Now for Sale: Hacking-as-a-Service', *Security Awareness*, McAfee, 27 December 2012, http://www.mcafee.com/uk/security-awareness/articles/now-for-sale-hacking-as-a-service.aspx.
16. The term 'phishing' was coined in the 1990s by hackers who were stealing America Online (AOL) accounts by scamming passwords from unsuspecting AOL users. See Gunter Ollmann, *The Phishing Guide: Understanding and Preventing Phishing Attacks* (Somers, NY: IBM Internet Security Systems, 2007), p.3.
17. 'Email Attacks: This Time It's Personal', Cisco Security White Paper, June 2011, http://www.cisco.com/en/US/prod/collateral/vpndevc/ps10128/ps10339/ps10354/targeted_attacks.pdf.
18. Ibid.
19. Gordon Corera, 'MI5 Fighting "Astonishing" Level of Cyber-Attacks', BBC, 25 June 2012, http://www.bbc.co.uk/news/uk-18586681.
20. 'Ethical Hacking', Version 5, Module II, Certified Ethical Hacker, http://dc347.4shared.com/doc/3ev5zbRv/preview.html.
21. See Zakir Durumeric, Eric Wustrow and J. Alex Halderman, 'ZMap: Fast Internet-Wide Scanning and its Security Applications', *Proceedings of the 22nd USENIX Security Symposium*, 14–16 August 2013, https://www.usenix.org/sites/default/files/sec13_full_proceedings.pdf.
22. For more information, see http://www.paterva.com/web6/.
23. See 'i2 Analyst's Notebook', IBM, http://www-03.ibm.com/software/products/gb/en/analysts-notebook/; and 'What we do', Palantir, http://www.palantir.com/about/.
24. See Johnny Long, *Google Hacking for Penetration Testers* (Burlington, MA: Syngress Publishing, 2008).
25. 'Google Hacking Diggity Project', Stach and Liu, http://www.stachliu.com/resources/tools/google-hacking-diggity-project/.
26. 'Cybercriminals Ditching Mass Spam for Targeted Attacks', Press Release, Cisco Technology News Site, 30 June 2011, http://newsroom.cisco.com/press-release-content?type=webcontent&articleId=422657.
27. 'Anatomy of an Attack', RSA Blog, April 2011, https://blogs.rsa.com/anatomy-of-an-attack/.

28. See, for example, www.ratemynetworkdiagram.com.
29. Mathew J. Schwartz, 'Anonymous Launches Operation Wall Street, Targets Ceos', *Informationweek Security*, 1 March 2013, http://www.informationweek.co.uk/security/attacks/anonymous-launches-operation-wall-street/240149804.
30. Stefano Baldi, Eduardo Gelbstein and Jovan Kurbalija, *Information Security and Organisations: A Non-Technical Guide to Players Offences and Effective Defences* (Geneva: DiploFoundation, 2003), p.37.
31. See, for example, Chris Boudreaux, 'Social Media Governance', *Policy Database, 2009–2013*, http://socialmediagovernance.com/policies.php.
32. Tim Bradshaw, 'Careless Tweets Cost Lives, MOD Warns', *Financial Times*, 14 June 2011.
33. 'MI6 Boss in Facebook Entry Row', *BBC News*, 5 July 2009, http://news.bbc.co.uk/1/hi/8134807.stm.
34. 'The Economist Explains: What Is the Streisand Effect?', *The Economist*, 15 April 2013, http://www.economist.com/blogs/economist-explains/2013/04/economist-explains-what-streisand-effect.
35. Ibid.
36. Matt Warman, 'Digital Right "to be Forgotten" will be Made EU Law', *The Telegraph*, 25 January 2012, http://www.telegraph.co.uk/technology/news/9038589/Digital-right-to-be-forgotten-will-be-made-EU-law.html.
37. Chris Paulus, 'CISPA 2.0: Say Goodbye to Our Constitutional Rights', Occupy.com, 28 February 2013, http://www.occupy.com/article/cispa-20-say-goodbye-our-constitutional-rights.
38. Megan Rose Dickey, 'Thanks To Apple and Google, Wearable Technology Is on Track to Become A $50 Billion Market', *Business Insider*, 19 May 2013, http://www.businessinsider.com/wearable-technology-market-2013-5.
39. Jeremy Hsu, ' "Smart Clothing" Could Become New Wearable Gadgets', *Live Science*, 1 February 2012, http://www.livescience.com/18238-smart-clothing-wearable-gadgets.html.
40. Parmy Olson, 'First-Known Targeted Malware Attack on Android Phones Steals Contacts and Text Messages', *Forbes*, 26 March 2013, http://www.forbes.com/sites/parmyolson/2013/03/26/first-known-targeted-malware-attack-on-android-phones-steals-contacts-and-text-messages/.
41. 'Google Glass Hacked Within Days Of Release', *Hackers News Bulletin*, 26 April 2013, http://www.hackersnewsbulletin.com/2013/04/google-glass-hacked-within-days-of.html?m=0.
42. 'Google Fined Over Illegal Wi-Fi Data Capture In Germany', *BBC News*, 22 April 2013, http://www.bbc.co.uk/news/technology-22252506.
43. 'Xbox One Kinect Is a surveillance Device and Anyone Can Get Remote Access to That', May 2013, http://hackersnewsbulletin.com/2013/05/xbox-one-kinect-is-surveillance-device.html.
44. Dave Evans, 'The Internet of Things: How the Next Evolution of the Internet is Changing Everything', *Cisco White Paper*, April 2011, http://www.cisco.com/web/about/ac79/docs/innov/IoT_IBSG_0411FINAL.pdf.

Part II
Open Source Intelligence and Proliferation

4
Armchair Safeguards: The Role of Open Source Intelligence in Nuclear Proliferation Analysis

Christopher Hobbs and Matthew Moran

Since the late 1960s, interest in preventing the spread of nuclear weapons has taken shape in 'an international regime based on commitment to the presumption of non-proliferation'.[1] Underpinned by the 1968 Treaty on the Non-proliferation of Nuclear Weapons (NPT), this collection of principles, norms, rules and processes has framed international action in the nuclear arena and nuclear proliferation has been characterised as 'deviant behaviour'.[2] Consequently, determining whether a state has engaged in proliferation in the past, is currently engaged in proliferation or has a propensity to proliferate in the future is intrinsically challenging due to a multitude of factors. These include the high levels of secrecy and compartmentalisation that surround nuclear weapons programmes and the difficulty in gauging the veracity of political statements regarding nuclear intentions. The 'dual-use' nature of much sensitive nuclear technology is also problematic, providing aspiring proliferants with a means of cloaking a weapons programme with a credible civil rationale, for a time at least.[3]

In this context, intelligence plays a key role in providing the technical information and knowledge necessary to inform assessments of the nuclear weapons ambitions and capabilities of state actors. However, the historical record of the intelligence community (broadly defined) in predicting and detecting proliferation is mixed, constituting a story of both high-profile successes and failures. The eventual success of intelligence efforts in unravelling the A. Q. Khan proliferation network, for example, was offset by the fact that Khan's export activities had gone on for many years.[4] Another commonly cited intelligence failure is provided by the US intelligence community's failure to predict India's nuclear tests in both 1974 and 1998, despite a number of indicators.[5]

In predicting nuclear proliferation, analysts have traditionally relied on information gathered from 'closed' or covert sources. However, while these methods can undoubtedly yield valuable information, they can also result in narrow assessments that rely excessively on specific and perhaps isolated sources. Considered without the benefit of context, these sources of information can misrepresent a situation and result in poor policy decisions. The intelligence failures associated with the war in Iraq should serve as a stark reminder in this regard.[6] This is not to suggest that open sources are impervious to this problem. Rather, we argue that the most accurate and valuable intelligence comes from a careful synthesis and analysis of data from the broadest possible range of sources. Analyses of data (of any sort) in isolation, and/or under the influence of political agendas, have led to intelligence failures in the past. The increasing availability of open sources provides the opportunity to add context and sometimes detail to other sources of information, including more traditional closed sources.

To gain an insight into a state's proliferation intentions (if any), it is thus necessary to evaluate a range of contextual, strategic, historical and technical information. Only with this broad understanding can individual pieces of intelligence be properly interpreted and understood. In this regard, OSINT has come to play a central role in intelligence efforts. Recent years have been marked by a tendency towards information-sharing, particularly online, and this has opened up a wealth of new possibilities for the intelligence community. It is now possible for a desk-based analyst to map, in general terms, a country's interests and activities in the nuclear domain. Indeed, as pointed out elsewhere in this volume, many senior figures in the intelligence community have claimed that upwards of 80 per cent of intelligence needs can be met by OSINT.[7]

This chapter will explore the use and value of open source information in nuclear proliferation assessments and, on a larger scale, how OSINT forms an integral part of what is termed 'all-source intelligence'. The focus will be on the work of the IAEA, whose analysts have, in their application of safeguards, adopted transparent, internationally accountable methods of data-gathering and analysis.[8] The analysis will show that open sources can yield new and important insights and we argue that OSINT has an increasingly important role to play in proliferation assessments.

The challenge of predicting nuclear proliferation

Since the dawn of the nuclear age, governments, international organisations, NGOs, academics, the media and others have attempted to make predictions about the spread of nuclear weapons. In 1963, for example, US President J. F. Kennedy, influenced by arguments about technological determinism, envisioned a world with 'fifteen, twenty, or twenty-five nuclear weapons powers' within the next decade.[9] However, this period saw nuclear weapons tested by just two new states – China in 1964 and India in 1974 – as other technically capable states decided not to pursue these particular weapons of mass destruction (WMD). A range of theories have been proposed by scholars of international relations, to explain the reasons why these states proliferated and others exercised restraint. These scholars weigh the influence of various driving factors, from domestic politics to psychological factors, in their efforts to produce a robust conceptual framework that will fully account for the drivers underlying decisions to proliferate.[10] However, while these theoretical approaches have served to illuminate the complex dynamics of proliferation in past cases, the range and diversity of the cultural, political, social and economic influences that must be considered in each individual context mean that few analysts would claim that these theories could function as highly accurate predictive tools.[11]

Nuclear proliferation analysis is further complicated by the challenge of obtaining accurate data upon which to base assessments. As mentioned, nuclear weapons programmes are typified by high levels of secrecy surrounding both the decision-making process and the necessary sensitive technical work, with only a handful of individuals involved in either aspect. For example, the UK's decision to pursue nuclear weapons was taken in 1947 by a specially convened cabinet subcommittee (Gen.163) consisting of then Prime Minister Clement Atlee and just five cabinet ministers.[12] In the case of India, between 1967 and 1974, no more than 75 scientists and engineers were involved in the weaponisation and detonation of nuclear explosives.[13] This level of compartmentalisation serves to limit the amount of directly relevant information available to the proliferation analyst attempting to accurately predict or lay bare the nuclear trajectory of a particular state.

The geographical expansion of the nuclear industry over the past 40 years together with the intrinsic dual-use nature of nuclear technology also serves to complicate nuclear proliferation analysis. States that develop uranium-enrichment capabilities for fuel production, or spent-fuel-reprocessing facilities for civil purposes, are technically not very far from being able to produce nuclear weapons, capable as they are of generating significant quantities of fissile material which could then be weaponised. This latent capability has led countries such as Japan, which has large-scale reprocessing facilities and stockpiles of separated plutonium, to be described as 'virtual' nuclear weapons states.[14] This despite the fact that Japan, with its particular history as the subject of nuclear weapons attacks, is seen as 'a champion of non-proliferation and disarmament'.[15] Consequently it can be difficult to judge whether the development of such technology is being pursued for purely civil purposes, as cover for a nuclear weapons programme or as part of a national strategy of 'nuclear hedging', which is defined by Levite as 'a national strategy of maintaining, or at least appearing to maintain, a viable option for the relatively rapid acquisition of nuclear weapons, based on an indigenous technical capacity to produce them within a relatively short time frame ranging from several weeks to a few years'.[16]

Proliferation-assessment strategies: The relevance of open sources

Although attempting to unpick a state's nuclear trajectory is a challenging task, a range of strategies can be employed to assist in evaluating the nature of a country's nuclear intentions. The utility of these strategies will inevitably vary from context to context, but common questions for which answers might be sought include: Has the political decision to pursue a nuclear weapons programme been taken? Technically, how far is a state away from producing nuclear weapons? Are a state's stated nuclear policies and activities in line with its obligations under international agreements and treaties? If a state were to decide to develop nuclear weapons, in what areas would it have to increase its technical capabilities, and what expertise and technology would it require to do so? Accurate answers to these and other questions are crucial in shaping the response of the international community. For example, an assessment might find that a state is likely to be pursuing a nuclear weapons programme but incapable of indigenously manufacturing essential nuclear warhead components. In this case, counterproliferation efforts could be directed at preventing the acquisition of the necessary technology – for example, through targeted technology-based sanctions and focused interdiction efforts.

In terms of analysing likely routes to nuclear weapons, gaps in development and suspicious nuclear activities, it is important to identify all of the technically plausible routes through which states could potentially acquire nuclear weapons. Different state technical capabilities may then be compared with this roadmap, in an approach commonly referred to as acquisition pathway analysis. Here, information of relevance would include a state's past, current and planned fuel cycle and research activities, with data gathered about facilities; materials; technical expertise; manufacturing base; and the ability to import relevant technologies.[17] The potential utility of this approach is highlighted by the US's inaccurate predictions regarding China's nuclear weapons development in the 1960s. Analysts judged 'that the Chinese would follow the United States and Soviet Union in what was seen as the technologically simpler plutonium path'.[18] However, the failure to consider alternative nuclear weapons acquisition routes meant that the US was taken by surprise by the fact that the weapon used in China's 1964 test was based on a fissile core of highly enriched uranium rather than plutonium.

However, an understanding of the various technical routes to the bomb is not enough; this must be accompanied by an assessment of potential political and strategic motivations. Stressing the importance of the manner in which a state positions itself below the nuclear threshold, for example, Quester notes that 'some states might elect to allow plutonium to accumulate first, before designing and building any models or prototypes of bombs; this would allow honest disclaimers that no "work on bombs" was underway'.[19] In this respect he emphasises that 'political capacities for violating treaties' must be measured 'as carefully as physical capabilities'.[20] The political and strategic decision-making calculus regarding the benefits of engaging in the pursuit of nuclear weapons must thus be carefully considered and weighed.

In general terms, then, assessing whether or not a state will make the decision to pursue a nuclear weapons programme will commonly take into account a range of variables, including their security situation (both internal and external); level of economic interdependence; understanding and representation of national identity; domestic political context; role of epistemic communities; technical capability; and availability (real or potential) of fissile material. A number of both qualitative and quantitative frameworks have been developed to weight the strength of the aforementioned factors.[21] For example, Bayesian Network models have been used to combine both social and technical factors in a quantitative manner, utilising assessment tools such as correlation, cluster and principle component analysis.[22]

Crucially, however, all of these strategies rely on accurate and credible information from inside the potentially nuclearising state. Where, then, is this information to be found? There is little prospect of those seeking to analyse nuclear behaviour obtaining a 'smoking gun' from open source information – that is to say, a direct insight into the processes underlying high-level political nuclear weapons decision-making processes and highly sensitive weaponisation work conducted by a state. However, OSINT can provide much contextual information and reveal strong indicators regarding a state's nuclear trajectory. Historical cases of proliferation allow us to deconstruct past efforts, successful or otherwise, to acquire the bomb. This can serve as a useful starting point, equipping researchers with an analytical framework within which current cases of concern can be considered. Useful information in this regard can be gleaned from government testimony, declassified documents, freedom-of-information requests and in-depth research studies.

Moving from the past to the present, there is a rapidly expanding range of publicly available sources that can shed light on the technical, political and strategic developments that often characterise the behaviour of those with nuclear weapons aspirations. The increasingly pervasive nature of the Internet in society, combined with the emergence of new forms of information generation, such as social media, means that open source data are more available and accessible than ever before, thus increasing their value as a source for analysts. Take the area of technical development, for example. Much information about the speed and direction of a state's technical progress can be gleaned from academic publications, scientific databases, patent records and satellite imagery, to name but a few sources. As is explored in Chapter 5, information about a state's illicit procurement efforts and technological capabilities drawn from the private sector, industry and business-to-business websites and trade publications can also be useful.

OSINT can provide information about the political and strategic aspects as well; indeed, from media reports to official statements, there is a substantial range of publicly accessible sources that can provide a wealth of information to support proliferation analysis in the nuclear arena.

The IAEA, nuclear safeguards and open sources

To demonstrate the contribution of OSINT to proliferation assessment, the following sections will focus on the identification, collection, synthesis and analysis of open sources conducted within the Department of Safeguards at the IAEA. The methods employed by the agency in monitoring safeguards are similar to those employed by proliferation analysts working in other organisations and domains. However, it is important to note that the approach of the IAEA differs significantly from that of national governments: as an independent international organisation, beholden to international law and comprising employees of various different nationalities, the IAEA cannot directly engage in the covert information-collection methods employed by national intelligence agencies. This makes the organisation relatively dependent on open source efforts. The IAEA is also unique in that it is empowered by Article III of the NPT to verify that non-nuclear weapons states (NNWS) are not diverting 'nuclear energy from peaceful uses to nuclear weapons'.[23] Furthermore, it is worth noting that unlike some other organisations, the IAEA does not seek to predict behaviour or intent, but focuses on past and current activities and capabilities.

Under the NPT, NNWS are required to sign a Comprehensive Safeguards Agreement (CSA) with the IAEA within 180 days of signing the treaty. A CSA obliges states to provide the IAEA with detailed records of nuclear (fissile) material inventories and flows, and permits IAEA inspectors to visit in-country nuclear facilities to gather data for use in verifying that nuclear activities are peaceful. This information is complemented by the IAEA's own open source research into country activities. The agency is thus 'in the unique position of being able to combine and compare State-declared information and inspection reports and findings, with information collected from a wide variety of open sources'.[24]

The advantage of drawing on the IAEA as a means of exploring the relationship between proliferation analysis and OSINT is that although both the specific data provided by national governments to the IAEA, and subsequent analyses of these data, are confidential, the organisation's status as the nuclear watchdog of the United Nations (UN) means that the methods used in this analytical process must be transparent and stand up to scrutiny. Consequently, there exists a significant body of readily accessible information that can be drawn upon to assess the IAEA's use of open source information.

The manner in which the IAEA both applies safeguards and utilises open source information in this process has evolved substantially over the years. In 1992, for example, following its failure to identify Iraq's clandestine nuclear weapons programme in the early 1990s, the IAEA Board of Governors decided that a CSA should not be limited to a state's declared nuclear material but should instead encompass all relevant nuclear material in a state. This decision altered the task of IAEA inspectors and analysts, and the objective of safeguards was broadened from verifying the 'correctness' of a state's nuclear declaration to ensuring its 'completeness', effectively transforming safeguards inspectors from nuclear material 'accountants' into 'detectives'.[25]

The Additional Protocol (AP), launched in 1997 as a measure that could be adopted in addition to a state's safeguards agreement, facilitated the implementation of this new approach by strengthening the IAEA's ability to detect undeclared nuclear weapons activity. Under the provisions of the AP, the level of access to nuclear facilities provided to IAEA inspectors is increased and technical verification methods are enhanced. Additionally, states are required to provide the IAEA with a far wider range of information about their nuclear fuel cycle activities, together with data about research and development work and nuclear-related imports and exports. The AP has been signed and implemented by the large majority of states. However, a number of exceptions remain. Iran, for example, has signed but not ratified the AP.[26]

Paradoxically, however, the increase in information required from state parties to the NPT 'fuelled the [IAEA's] requirement for open source research' as the agency sought additional means of verifying a greatly expanded 'array of locations and activities'.[27] Although there is no direct reference to the exploitation of open sources within the AP, it was around this period that 'comprehensive capabilities' for the collection and processing of open source information were established in the Division of Safeguards Information Management (SGIM), one of six divisions and two offices that make up the Department of Safeguards.[28]

The task of evaluating a state's nuclear activities and, in particular, verifying the absence of undeclared nuclear material and activities is a challenging one. In tackling this task, the IAEA adopts a 'state-level approach' (SLA) in which a range of disparate information (including state-supplied reports, data from IAEA inspections, and third-party and open source information) is collected and analysed. This is an iterative and 'information-driven' process, with conclusions forming the basis of plans for further, targeted safeguards activities.[29] What constitutes safeguards-relevant information is determined through acquisition pathway analysis based on the IAEA 'Physical Model' of the nuclear fuel cycle and weapons-development processes.[30] This allows a range of indicators (including relevant equipment, nuclear material and non-nuclear material) relating to all plausible processes for converting civil nuclear materials into material of potential use in weapons to be identified.[31] Detail is crucial here. For example, the type and grade of carbon fibre being produced in a country can be revealing when placed in a particular context of nuclear development.

Open source information plays an important role as part of the all-source analysis process of the SLA. In general terms, open sources, correctly interpreted and understood, can provide information about political and economic developments, international relations and security perceptions, offering the IAEA an insight into the broader context in which a government's decisions related to nuclear technology are taken.[32] At a more operational level, OSINT can be used to support the verification of a state's declarations by providing, for example, technical details about facilities or equipment, details regarding nuclear-relevant industrial infrastructure, information about the capabilities of research laboratories and an understanding of the historical context surrounding a state's nuclear programme. Open source information in the form of commercial satellite imagery sometimes also serves as the IAEA's sole source of data for monitoring developments at unsafeguarded sites and facilities.

Furthermore, with the nuclear industry, like so many others, increasingly operating across borders, with vendors from several countries typically involved in large projects, open source information helps the IAEA to look beyond the complexities of individual states and better understand the links connecting multiple states' nuclear facilities and nuclear-relevant commercial manufacturers.[33] Open sources are also synthesised with IAEA verification data and third-party information in confirming that state declarations are complete and correct. For example, open source trade data can be used to help to verify that a state's declarations of nuclear-related imports and exports are accurate under its AP. Moving beyond the SLA, the IAEA also uses open sources to analyse transnational proliferation networks. This aspect of the IAEA's work is discussed in detail in Chapter 5.

OSINT challenges and the IAEA strategy for exploiting open sources

The described evolution of the IAEA safeguards system coincided within the information revolution of the late twentieth century, where a series of technological advancements in computing and digital communications served to dramatically increase the global capacity to store, generate and exchange information.[34] This has resulted in an 'overwhelming and growing quantity of information available on all thematic areas of relevance to (IAEA) safeguards'.[35] In addition to this enormous proliferation of information, safeguards analysts must also cope with information in multiple languages, often of questionable reliability and in differing and sometimes unstructured formats (for example, raw technical data and satellite imagery usually require specialist interpretation). In short, the IAEA is faced with an acute case of information overload, a problem compounded by the need to separate the wheat from the chaff and draw actionable conclusions from all of the information gathered. The situation today is thus far removed from the IAEA's original task – the collection and analysis of structured, state-provided nuclear material accounting data – and the Safeguards Information Management team (and, indeed, the Department of Safeguards as a whole) has had to evolve rapidly to meet this challenge, through the adoption of new methodologies and workflows, the creation of specialists databases, the employment of OSINT experts and the implementation of technical tools.

The collection and analysis of open source information for the Department of Safeguards is based upon the continuous monitoring of a diverse range of sources for information that could be of safeguards relevance with respect to individual or multiple states. Information deemed to be of relevance is gathered and categorised according to state or topic, before being subjected to a process of review and analysis. The products of this work are provided to various constituencies of interest within the IAEA, but most open source assessments are provided to the relevant State Evaluation Group (SEG) along with the information gleaned from all of the other forms of information collection at the agency's disposal.[36] These small, multidisciplinary groups of staff work collaboratively and on an ongoing basis to evaluate each member state's nuclear activities with respect to its safeguards obligations.[37]

Clearly, then, the success of the IAEA depends on resilient flows of accurate information that reach the right people at the right time. In this context, a major programme is currently under way at the IAEA to implement a new secure, collaborative virtual work environment known as the Integrated Safeguards Environment (ISE). This is intended to enhance the agency's information acquisition and analysis functions by streamlining information-sharing in an appropriately secure manner. More specifically, it 'has created the architecture and the information environment to enable all-source analysis in the Department of Safeguards, particularly with the fusion of all key data sources and the improvement of the key business processes in collecting data from Member States'.[38]

The IAEA has adopted a number of distinct approaches for the collection, analysis and distribution of open source information. To keep track of new and ongoing developments in nuclear proliferation and safeguards issues, IAEA open source analysts employ automated tools in combination with 'refined search procedures' to filter thousands of information items from a variety of open sources, including trade publications and specialist databases.[39] Specialist software is also used to 'locate and collect information that has not been indexed'.[40] The IAEA's reach here is wide, from the broader, more generic searches for safeguards-relevant information to the targeted searches that aim to address a specific safeguards issue in a particular state, to help to evaluate and verify a state's AP declarations, to support onsite inspection activities, to gain additional insights into the research and development work being undertaken by a state in a particular field, or to investigate allegations of illicit procurement activities.[41]

The Department of Safeguards conducts significant amounts of research and analysis internally. However, as part of its response to the constraints faced by the IAEA in terms of resources, combined with the ever-increasing amounts of information available online, it has also turned to external experts for some of its open source collection work. This collaboration boosts both the efficiency and the reach of the Agency's information-collection machinery. The strong growth in foreign language sources, for example, demonstrates the need for the IAEA to utilise the 'linguistic skills of highly trained collectors and analysts'.[42] Consequently, the IAEA has established Regional Information Collection Centres (RICC) at King's College London aimed at focusing OSINT expertise for this very purpose. This initiative is funded as part of the UK Safeguards Support Programme and has 'extended the Agency's ability to identify relevant information, without which the Agency's confidence in safeguards conclusions would be reduced'.[43]

Once gathered, information is then distributed through multiple channels within the IAEA. Information of broad relevance in the context of safeguards is sent to departmental staff, via an email newsletter: the 'SGIM Open Source Daily Highlights'. More narrowly focused items of information are emailed directly to particular staff members according to their specific work assignments and areas of interest. In addition to being immediately brought to the attention of safeguards staff, open source information is categorised via a nuclear fuel cycle 'topic tree' structure and fed into the Safeguards Department's central open source database. Analysts are then able to access this information at a later date via an internal, free-text search engine.

At the edge of practice: OSINT and proliferation case studies

Clearly, open source information plays an important role in the safeguards process on a daily basis, both directly, by helping to confirm that a state's declaration to the IAEA is accurate and verifiable, and indirectly, by providing a contextual lens through which a state's nuclear programme can be examined. However, these ongoing 'positive' effects of OSINT in the safeguards process are rarely given attention in the public domain. Rather, it is the cases where open source information has helped the IAEA to identify potential proliferation issues that attract interest and attention, for obvious reasons. Drawing on the cases of A. Q. Khan's nuclear 'black market' network in Egypt and Libya, for example, Wyn Bowen demonstrates the value that may be derived from OSINT in practice.[44]

The case of Egypt is particularly interesting since information gleaned from open sources served as a 'tip-off' to the IAEA in 2004, when technical documents published by current and former staff at the Egyptian Atomic Energy Authority revealed information that suggested that Cairo's safeguards declaration did not fully reflect the country's nuclear activities. The insights gleaned from open sources prompted an IAEA visit to the Inshas Nuclear Research Centre north of Cairo, where it was confirmed that Egypt had failed to report a range of research and development activities involving small amounts of nuclear material.[45]

Open source information was also one of the types of data upon which the IAEA drew to support its efforts to verify South Korea's state declaration following the entry into force of the AP to its safeguards agreement in February 2004. This case differed from the Egyptian one in that the government of South Korea informed the IAEA of previously undeclared experiments dating from 2000.[46] This belated declaration prompted an IAEA inquiry. An inspection team was sent to South Korea to verify the declaration, while in Vienna the Department of Safeguards sifted through accumulated information relating to the Korean programme. In this context, information obtained from open sources led to a further disclosure from Seoul regarding an additional, undeclared uranium chemical enrichment experiment carried out by South Korean scientists during the period 1979–1981.[47] As in the Egyptian case, open source information provided the basis for an IAEA request for clarification, which, in turn, prompted South Korea to provide information about a project that should have been reported previously.

OSINT has also played an important role for the IAEA in assessments of Iran's nuclear activities over the last decade. In August 2002, for example, the IAEA responded to media reports that Iran was building an undeclared underground nuclear facility at Natanz and a heavy water production plant at Arak by requesting (and subsequently obtaining) information about Tehran's nuclear fuel cycle development plans.[48] Open sources also provided information that resulted in the Iranian authorities admitting that uranium enrichment-related activities had been conducted at the workshop of the Kalaye Electric Company in Tehran.[49] Later, a range of open source information was also used by the IAEA in successful efforts to identify the Pakistani origin of Iranian centrifuge components contaminated with traces of highly enriched uranium. Here, targeted open source searches were used to identify 'enrichment research or nuclear cooperation with another State that has mastered enrichment technology' and to 'determine the past and present nature of any nuclear cooperation, including the technical capabilities of the exporter'.[50] More recently the IAEA has used open sources to shed light on Iran's weaponisation capability, with scientific publications revealing that Iranian scientists have carried out work on both neutron transport and explosive modelling.[51] Commercially available open source satellite imagery has also been used to remotely monitor activities at Iran's heavy water production plant in Arak and the Parchin military complex, sites which have not, at the time of writing, been opened up to IAEA inspection.[52]

In contrast with the successful application of open source information in the cases of Egypt, South Korea and Iran, retrospective analysis of both Libya's and India's nuclear weapons programmes demonstrates how in these cases open sources would likely have only provided 'contradictory' or 'mixed' indicators in assessing proliferation behaviour.[53] In the case of Libya, an examination of open sources in the late 1990s would likely have concluded that Gadhafi had decided to end his pursuit of WMD. This period saw the Libyan regime sign up to the Hague Code of Conduct on Ballistic Missile Proliferation and 'move significantly away from its past involvement in sponsoring terrorism, attempts to undermine other governments in the region and traditional bellicosity towards Israel'.[54] However, although Libya did ultimately give up its nuclear weapons ambitions, it actively pursued a programme up until October 2003, acquiring centrifuge technology through the A. Q. Khan network, and only deciding in December of that year to abandon its nuclear weapons aspirations.

In the case of India, an analysis of publically available political statements preceding its 1998 nuclear tests would have provided proliferation analysts with mixed signals regarding Delhi's nuclear intentions. Although the Bharatiya Janata Party (BJP) declared its intention to 'exercise [India's] option to induct nuclear weapons' prior to the election of Behari Vajpayee as Prime Minister, statements made after the election took a softer approach, with the party toning down its pro-nuclear weapons rhetoric.[55]

Both the Libyan and the Indian cases thus demonstrate why good intelligence requires the assessment and synthesis of all available information and sources. As with covert sources of intelligence, the analysis of open sources in isolation can be problematic.

Conclusions

This chapter has sought to show how OSINT can shed light on an arena typically characterised by secrecy, subterfuge and covert action. When exploited as part of a broader analytical process that examines and cross-correlates many data streams, open source information can provide valuable insights into the proliferation-relevant activities of states aspiring to nuclear weapons. Two key benefits stand out in this respect. First, open sources can provide valuable contextual information within which a state's nuclear decision-making and technical capabilities can be assessed. By adopting a collection strategy that is both wide-ranging and comprehensive, information can be gathered on a number of potential indicators – political, diplomatic and technical – all of which expand and embellish the IAEA's understanding of a particular state's nuclear programme. Second, open source information can serve to direct specific lines of inquiry and research, providing a 'tip-off' function that can focus more technical means of inspection and verification. In the case studies referred to above, open source information has triggered investigations that have found states to be in non-compliance with their safeguards agreements. Thus while OSINT will rarely (if ever) provide a 'smoking gun', the use and value of open source information should not be underestimated.

On a larger scale, it is important to consider OSINT in the context of the structural challenges faced by the IAEA. As more and more countries turn to nuclear power as a means of answering growing energy needs, the demands placed on the agency continue to grow in proportion. Moreover, the current, inhospitable economic climate means that demand vastly outstrips supply in terms of the resources at the agency's disposal. The IAEA is implicitly being asked to expand its work without a concomitant expansion of its resources. This means that efficiency and effectiveness must form the cornerstones of the IAEA's approach. And as a relatively cost-effective means of gathering information, open source research plays a crucial role in the IAEA's activities. Developments in OSINT tools, techniques and methodologies, some of which are discussed in this volume, combined with the vast and ever-increasing amount of information available online, mean that open sources can potentially provide significant benefits for little additional cost. In general terms, then, it seems clear that the exploitation of open sources will occupy an increasingly important role in the strategic approach of the Department of Safeguards at the IAEA.

Notes

1. Avner Cohen and Benjamin Frankel, 'Opaque Nuclear Proliferation', *Journal of Strategic Studies* (1990), Vol. 13, No. 3, p.16.
2. Ibid., p.17.
3. Dual-use technology refers to products and technologies normally used for civilian purposes but which may have military applications.
4. For a detailed account, see David Albright, *Peddling Peril: How the Secret Nuclear Trade Arms America's Enemies* (New York: Free-Press, 2010).
5. Torrey C. Froscher, 'Anticipating Nuclear Proliferation: Insights from the Past', *The Nonproliferation Review* (2006), Vol. 13, No. 3, pp.467–477.
6. For a comprehensive study of intelligence failures around the Iraq War, see Robert Jervis, *Why Intelligence Fails: Lessons From the Iranian Revolution and the Iraq War* (Ithaca, NY: Cornell University Press, 2010).
7. See Chapter 1.
8. According to the definition provided by the IAEA, safeguards comprise 'an extensive set of technical measures by which the IAEA Secretariat independently verifies the correctness and the completeness of the declarations made by States about their nuclear material and activities'. See 'What We Do: Safeguards', website of the IAEA, http://www.iaea.org/safeguards/what.html.
9. James E. Doyle (ed.) *Nuclear Safeguards, Security and Non-proliferation: Achieving Security with Technology and Policy* (Oxford: Elsevier, 2008), p.21.
10. Tanya Ogilvie-White, 'Is there a Theory of Nuclear Proliferation? An Analysis of the Contemporary Debate', *The Nonproliferation Review* (1996), Vol. 4, No. 1, pp.43–60.
11. Ibid.

12. Margaret Gowing, *Independence and Deterrence: Britain and Atomic Energy, 1945–1952. Volume 2: Policy Execution* (London: Macmillan, 1974), p.442.
13. George Perkovich, *India's Nuclear Bomb: The Impact on Global Proliferation* (California: University of California Press, 1999) p.172.
14. Aurelia George Mulgan, 'Why Japan Still Matters', *Asia-Pacific Review* (2005), Vol. 12, No. 2, p.108.
15. Ariel E. Levite, 'Never Say Never Again: Nuclear Reversal Revisited', *International Security* (2002), Vol. 27, No. 3, p.71.
16. Ibid., p.69.
17. See Jill N. Cooley, 'Progress in Evolving the State-Level Concept', paper presented at the Seventh INMM/ESARDA Joint Workshop Future Directions for Nuclear Safeguards and Verification, Aix-en-Provence, France (17–20 October 2011).
18. Torrey C. Froscher, 'Anticipating Nuclear Proliferation Insights from the Past', *Nonproliferation Review*, Vol. 13, No. 3, November 2006, p.470.
19. George H. Quester, 'Some Conceptual Problems in Nuclear Proliferation', *The American Political Science Review* (1972), Vol. 66, No. 2, p.493.
20. Ibid., p.497.
21. Alexander H. Montgomery and Scott D. Sagan, 'The Perils of Predicting Proliferation', *Journal of Conflict Resolution* (2009), Vol. 53, No. 2, pp.302–328.
22. G. A. Coles et al., 'Utility of Social Modelling in Assessment of a State's Propensity for Nuclear Proliferation', *Pacific Northwest National Laboratory Report, PNNL-20492*, June 2011, http://www.pnnl.gov/main/publications/external/technical_reports/PNNL-20492.pdf.
23. United Nations, 'The Treaty on The Non-Proliferation of Nuclear Weapons (NPT)', http://www.un.org/en/conf/npt/2005/npttreaty.html.
24. Michael Barletta, Nicholas Zarimpas and Ryszard Zarucki, 'Open Source Information Acquisition, Analysis and Integration in the IAEA Department of Safeguards', paper presented at the James Martin Center for Nonproliferation Studies, Twentieth Anniversary Celebration: The Power and Promise of Nonproliferation Education and Training 3–5 December 2009, http://cns.miis.edu/activities/20th_anniversary/media/presentation_paper_barletta_zarimpas_zarucki.pdf.
25. Theodore Hirsch, 'The Additional Protocol: What it is and Why it Matters', *The Nonproliferation Review* (2004), Vol. 11, No. 3, p.143.
26. IAEA, 'Conclusion of Additional Protocols: Status as of 24 September 2013', IAEA website, September 2013, http://www.iaea.org/safeguards/documents/AP_status_list.pdf.
27. Wyn Q. Bowen, 'Open Source Intelligence and Nuclear Safeguards', in Robert Dover and Michael S. Goodman (eds.), *Spinning Intelligence: Why Intelligence Needs the Media, Why the Media Needs Intelligence* (New York: Columbia University Press, 2009), p.94.
28. Barletta, Zarimpas and Zarucki, 'Open Source Information Acquisition, Analysis and Integration in the IAEA Department of Safeguards'.
29. IAEA, 'The Safeguards System of the International Atomic Energy Agency', *IAEA Document*, p.8, www.iaea.org/safeguards/documents/safeg_system.pdf?.
30. The Physical Model is an analytical tool that 'Identifies, Describes and Characterises Fuel Cycle Technologies and Processes'. See Annette Berriman, Russell Leslie and John Carlson, 'Information Analysis for IAEA Safeguards', paper presented at the Institute of Nuclear Materials Management (INMM) Symposium, 18–22 July 2004, p.2, http://www.dfat.gov.au/asno/publications/inmm2004_information.pdf.

31. Ibid.
32. See Barletta, Zarimpas and Zarucki, 'Open Source Information Acquisition, Analysis and Integration in the IAEA Department of Safeguards'.
33. Ibid.
34. Martin Hilbert and Priscila Lopez, 'The World's Technological Capacity to Store, Communicate, and Compute Information', *Science* (2011), Vol. 332, No. 6025, pp.60–65.
35. Barletta, Zarimpas and Zarucki, 'Open Source Information Acquisition, Analysis and Integration in the IAEA Department of Safeguards', p.2.
36. Ibid.
37. For an overview of how safeguards are conceptualised and implemented by the IAEA, see 'The Conceptualization and Development of Safeguards Implementation at the State Level', report by the Director General, IAEA Board of Governors Report, GOV/2013/38, 12 August 2013, http://armscontrollaw.files.wordpress.com/2012/06/state-level-safeguards-concept-report-august-2013.pdf. For an overview of how open sources are incorporated into the work of the Department of Safeguards, see 'Development and Implementation Support Programme for Nuclear Verification 2012–2013 (STR-371)', *International Atomic Energy Agency*, 2011, http://www.bnl.gov/ispo/docs/pdf/RD%20Programme/D-IS_ProgrammeForNuclearVerification_2012-2013.pdf.
38. 'Development and Implementation Support Programme for Nuclear Verification 2012–2013 (STR-371)', p.84.
39. Barletta, Zarimpas and Zarucki, 'Open Source Information Acquisition, Analysis and Integration in the IAEA Department of Safeguards', p.3.
40. Ibid., p.5.
41. Michael Barletta et al., 'Information Analysis for Additional Protocol Evaluation', *Presentation to the INMM Annual Meeting*, 18th–22nd July 2004.
42. Ibid.
43. J. W. A. Tushingham, 'UK Safeguards Support Programme: Report on Activities and Progress during the Period 1 April 2011 to 31 March 2012', Department of Energy and Climate Change document, August 2012, p.12, https://www.gov.uk/government/uploads/system/uploads/attachment_data/file/65519/6403-uk-safeguards-support-report-2011-12.pdf.
44. See Bowen, 'Open Source Intelligence and Nuclear Safeguards'.
45. 'Implementation of the NPT Safeguards Agreement in the Arab Republic of Egypt', *IAEA Board of Governors Report*, GOV/2005/9, 14 February 2005, http://www.globalsecurity.org/wmd/library/report/2005/egypt_iaea_gov-2005-9_14feb2005.pdf.
46. 'Implementation of the NPT Safeguards Agreement in the Republic of Korea', *IAEA Board of Governors Report*, GOV/2004/84, 11 November 2004, p.1, http://www.iaea.org/Publications/Documents/Board/2004/gov2004-84.pdf.
47. 'Implementation of the NPT Safeguards Agreement in the Republic of Korea', p.2, p.7.
48. 'Implementation of the NPT safeguards agreement in the Islamic Republic of Iran', *IAEA Board of Governors Report*, GOV/2003/40, 6 June 2003, p.2, http://www.iaea.org/Publications/Documents/Board/2003/gov2003-40.pdf.
49. Ibid.
50. L. Bevaart et al., 'Safeguards Information Analysis: Progress, Challenges and Solutions', Paper IAEA-CN-148/26, Proceedings of an International Safeguards Symposium on Addressing Verification Challenges, 16–20 October 2006, p.61,

http://www-pub.iaea.org/MTCD/publications/PDF/P1298/P1298_Contributed_Papers.pdf.
51. 'Implementation of the NPT Safeguards Agreement in the Islamic Republic of Iran', IAEA Board of Governors Report, GOV/2011/65, 8 November 2011, p.11, http://www.iaea.org/Publications/Documents/Board/2011/gov2011-65.pdf.
52. 'Implementation of the NPT Safeguards Agreement and Relevant Provisions of Security Council Resolutions in the Islamic Republic of Iran', IAEA Board of Governors Report, GOV/2013/6, 21 February 2013, http://www.iaea.org/Publications/Documents/Board/2013/gov2013-6.pdf.
53. Bowen, 'Open Source Intelligence and Nuclear Safeguards', p.102; Richard A. Best, Jr, and Alfred Cumming, 'Open Source Intelligence (OSINT): Issues for Congress', *CRS Report for Congress*, 5 December 2007, p.24, http://www.fas.org/sgp/crs/intel/RL34270.pdf.
54. Bowen, 'Open Source Intelligence and Nuclear Safeguards', p.101.
55. Howard Diamond, 'After BJP Election Win, Leaders Soften Line on Nuclear Weapons', *Arms Control Today*, March 1998, http://www.armscontrol.org/print/308; Perkovich, 'India's Nuclear Bomb: The Impact on Global Proliferation', p.408.

5
Open Source Intelligence and Proliferation Procurement: Combating Illicit Trade

Daniel Salisbury

The proliferation of unconventional weapons – nuclear, chemical and biological, and their means of delivery – to countries such as Iran and North Korea is recognised as one of most pressing international security issues today. Historically, states pursuing such programmes have benefitted from technology or assistance from outside their jurisdictions. The US nuclear programme, for example, benefitted from the expertise of émigré scientists from Europe and uranium from the Belgian Congo. In the present day, aspirant states do not have the advanced manufacturing or knowledge base to produce all of the required technology in sufficient quantities and qualities. Consequentially, these states have been seeking these technologies from the international marketplace.

Because of the international community's desire to prevent proliferation, a complex web of export controls, UN, EU, and national sanctions and embargoes have been put in place. This has led proliferators to acquire many of the prerequisite technologies by illicit procurement. Illicit procurement encompasses efforts to acquire these technologies in breach of both the letter and the spirit of these measures, often involving deceptive methods, such as the use of front or shell companies, falsifying documents, disguising shipments and using unconventional shipment routes. Efforts to prevent the acquisition of such technology can slow programmes down and allow time for diplomatic efforts to be made. The value of supply-side measures was highlighted by the chief of the British foreign intelligence service, MI6, who noted in 2012 that service efforts, which included 'a series of operations to ensure that the sanctions introduced internationally are implemented', had potentially set back Iranian nuclear progress by six years.[1]

While intelligence agencies and covertly acquired information has generally led in the fight to counter the proliferation of WMD, there is now

82 OSINT and Proliferation

a larger body of publically available and useful information relating to illicit procurement than ever before. However, much of this information is fragmented, relates to illicit transactions that have already occurred, and therefore has limits in preventing illicit transactions that may occur in the near future. Similarly, industry and government have sight of different types of open source information that may or may not be useful when considered alone. Despite these challenges, it is clear that the potential of open source in preventing illicit trade in the future is significant. Effective information collection, management and dissemination processes could provide for a fuller picture of illicit procurement activities to governments and an increased awareness of the risks to industry.

This chapter explores the extent and the limits of open source information, and its role in preventing illicit trade. It argues that while most of the value of open source is in providing background and contextual information, there is also a clear potential for open source to fulfil a real-time operational role, benefitting both governments and industry around the world. In this regard, the effective use of open sources is capable of having a serious impact on proliferation hard cases. This chapter will begin by considering some of the new sources of information available in the public domain before discussing the broader role that open source intelligence can play for states and the private sector. It will conclude by considering some of the practical challenges in using open sources in this way, the limitations of current open source processes, and some ways in which open sources could be harnessed in the future to most effectively prevent the illicit procurement of proliferation-sensitive goods.

Open source: Opportunities and challenges of a growing body of information

There is now more information relating to illicit procurement in open sources than ever before. The growth of this information has especially been seen in the past few years and can be attributed to a number of factors. The increased policy focus on preventing transfers of technology to Iran and North Korea has formed part of this. UN technology-based sanctions have been in place since 2006.[2] As well as putting barriers in place around the world to prevent these countries from openly purchasing goods, sanctions have raised awareness of procurement activities. News reports have increasingly detailed violations, alongside the UN Expert Panel reports, as well as court indictments and associated reporting as cases emerge. There has also been growth in secondary sources, which seek to provide analysis of this information. A number of NGOs, think tanks and academics frequently publish on illicit trade issues.[3]

A more specific series of events, which has impacted massively on the volume of available information, has been the leak of US diplomatic cables by the activist group Wikileaks. The organisation has released over 250,000 cables covering the period from 1966 to 2010, but mostly on illicit trade issues relating to the 2007–2010 period.[4] A quick search of one of the cable databases, for example, reveals 14,108 leaked cables including the word 'proliferation'.[5] Many of these detail communications between the US State Department and US embassies around the world relating to illicit trade. They often feature the names of entities implicated, technologies allegedly being transferred, and sometimes more detailed information relating to alleged shipping routes and proliferators' methods. While drawing extensively on more traditional 'closed' intelligence, all of these cables are, for better or for worse, now available in the public domain.

However, there are both general and specific limits to these multiple sources of information. More broadly, the nature of illicit trade means that those partaking in it often use deceptive methods to avoid discovery. Even when discovered, many continue to deny their involvement and are reluctant to elaborate on specifics.[6] Therefore individual documented cases are often lacking in specific verifiable detail. More broadly, when it comes to illicit trade of WMD-related goods, like other illicit activities such as narcotics smuggling or counterfeit goods trafficking, it is impossible to collect a complete dataset; beyond the 'collectable' data, there are an unknown number of unknown cases. This makes it difficult to contextualise the relative importance of various cases or pieces of information, or the prevalence of certain practices.

Different types of information have more specific limitations. If we take the cables leaked by Wikileaks as an example, they only detail the US side of the dialogue. Most cables sent from the State Department take the form of a demarche relating to an alleged transaction that the staff based in the US Embassy in a specific country is tasked to deliver to that country's foreign ministry. Cables sent back to Washington detailing the country's response vary in value depending on the country in question's attitude to non-proliferation, and also its relationship with the US. In some cases, an ongoing dialogue can be mapped out through multiple cables, and cross-referenced with other open sources. However, in others there is no further information beyond one or two cables.

Also, the nature of the leaked cables is important to note. The highest classification of the cables released is 'secret', with most of the cables being 'confidential' or 'unclassified'.[7] This has an effect on cable content, meaning that they mostly detail proliferation of missile, chemical, biological or military-related items rather than more sensitive nuclear-related goods. Also, there is, in most cases, no possibility of verifying the information that they contain. While details of the activities of some entities are to be found in other open sources, such as indictments or press reports, the majority cannot be found. Similarly, the information contained in Wikileaks and many other open sources, while providing useful background and past examples, is usually not timely enough to provide operational intelligence for governments and industry on the ground.

While deriving utility from a large proportion of the available open source information presents great difficulty, it still has a clear role to play for states and the private sector alike in preventing illicit transactions and meeting non-proliferation objectives. Both governments and private sector actors have access to different types of information, which can be of great benefit to the other in working to preventing illicit transfers. For private sector actors, open sources, by definition, are all that are available to them given that they lack access to the more traditional intelligence available to governments. Open sources in this respect can be used to overcome some of the sensitivities and contribute to an informed industry. Industry itself also possesses information from the coalface relating to illicit trade.

A valuable component of all-source: The utility of open sources for national authorities

States have access to intelligence sources of all types, which, through collection and analysis, feed into 'all-source' intelligence products. When it comes to nuclear proliferation and, more specifically, illicit trade, states obviously collect intelligence in a number of more 'traditional', classified ways. For example, multiple national intelligence agencies used various covert intelligence sources to uncover the A. Q. Khan network.[8]

However, more covert intelligence efforts have not always been shown to deliver in uncovering illicit procurement activities. Take, for example, the intelligence failure in Iraq in the early 1990s when the world's most sophisticated intelligence organisations failed to detect the country's activities.[9] Iraq's efforts involved widespread illicit procurement of technologies from the international marketplace. With this in mind, it is clear that the less conventional intelligence sources – open sources – can be of great use to states. The information that relates to 'suspicious enquiries' and technical capabilities, which can be gathered from the private sector, is considered below.

Suspicious enquiries: Information from beyond the licensing process

Significant quantities of information are obtained by states through their implementation of supply-side controls on sensitive technologies. In the present day, most states that have private sectors capable of manufacturing sensitive technologies have a system of export controls in place. These systems allow national authorities (the generic descriptor used to describe the various organs of a state involved in licensing decisions) to make a risk assessment of each individual sensitive export before it is made.

A number of efforts over the years have been made to internationally harmonise efforts to control such technologies. Many of the supplier states are either members of or utilise the lists and guidelines agreed upon by international supplier regimes, such as the Nuclear Suppliers Group (NSG), the Missile Technology Control Regime (MTCR) and the Australia Group. States also implement the technology-based sanctions listed in UN resolutions through their export control systems. Together these controls and sanctions regimes are designed to prevent goods from being exported for uses that are deemed by governments to be unacceptable: in human rights abuses, certain military programmes or build-ups, and WMD programmes.

Besides physically denying technology to programmes of concern, export controls also seek to deter states' interests in such programmes by increasing their difficulty in acquiring technology. Crucially, they also play a valuable detection role. Through the submission of licence applications, exporters are giving governments several pieces of information regarding the more contentious of various entities' attempts to purchase technology. Typical licence applications include information regarding the alleged end user, the technology (including technical specification) and the consignee.[10] Applications often include some kind of End User Undertaking (EUU), signed and completed by the alleged end user. EUUs often need to be attached to a covering letter from the end user or consignee, which includes an official letterhead and logo. The licence applications that are submitted by companies and later refused on WMD-related grounds can provide valuable intelligence. Also, if licences are found to have been granted in error, this information may allow governments to follow up if goods are found to have reached programmes of concern.

However, the export control system alone is insufficient in its collection of information about illicit procurement trends. This is due to the nature of export controls more generally, the deceptive methods used by proliferators, and the state-centric focus of control systems. In most export control systems, licences are only required in two broad cases: first, when the technology to be exported is listed on national control lists; and second, when an exporter 'knows or suspects' that technology may be destined for a WMD programme or there is a risk of diversion (even when the technology is not controlled). This creates a number of situations where a licence application is not necessary, and the national authority is not necessarily privy to the useful information regarding the export.

In order to circumvent export controls, those seeking to knowingly illicitly transfer or procure technology often operate in highly deceptive ways. Avoiding controls, they often deliberately seek non-controlled technologies that can be substituted for controlled technologies. The UN Panel of Experts has noted this with regard to Iranian illicit procurement activities.[11] Proliferators – both the elements of states that are seeking the capability, and private sector actors who are assisting them – are obviously not open about their intentions, often making it difficult for exporters to judge the risk of diversion. In fact, they may just not apply for licences at all in order to try to get goods through without alerting the national authority.

On the other hand, national authorities are often deprived of information because industry does not always apply for a licence or share information with government when they receive a business enquiry that they assess to be suspicious. When contacted regarding business that they believe to be clearly dubious, many firms will not apply for a licence or retain the details contained in the email, letter or telephone conversation. There may be other reasons beyond non-proliferation-related legal and financial risks for rejecting orders and not applying for a licence.[12] For example, a firm may not stock the product or may be unable to export it.

Even in the most obvious cases where enquiries are clearly from a sanctioned programme, there may be reasons why a firm does not share information. Perhaps they do not have an appropriate contact in government to pass the information onto. In some countries, such as the US, there is also concern in industry that passing information on may result in prosecution.[13]

The number of licence applications rejected by the national authorities tends to be low. Export licensing statistics for 2012 in the UK reveal that just 190 Standard Individual Export Licences (SIELs) were refused, with 12,910 being granted.[14] Also, individual governments have an incomplete picture given that the information gathered through the licensing process only relates to licence applications made within their jurisdiction, unless they are privy to licence refusal sharing with allies or through the export control regimes.

In short, while export controls are there, in part, to 'detect' and gather information about sensitive exports, the most valuable information is not gathered through this process. There is real value in information which exporters have sight of when they receive suspicious enquiries and would not normally pass to national authorities. As an industry representative noted, 'any piece of information that remains in our company archives or ends up in the trash bin is forever lost for government authorities'.[15] However, as will be discussed, there are significant difficulties for national authorities in motivating firms to go beyond compliance, and to retain and provide this information. These difficulties concern surrounding commercial sensitivities, available time and resources, and in some cases fear of prosecution or reputational damage.

Technical information: Assessing capabilities and possibilities

Open source information can also be of use to national authorities in assessing the technical capabilities of different actors to manufacture proliferation-sensitive technologies. This can be analysed alongside information about suspicious enquiries to build a picture about proliferators' illicit procurement activities allowing governments to assess the level to which a proliferating country's industry is self-sufficient, whether procurement from the international marketplace is necessary, and which firms overseas may be targeted for sensitive technologies. Some of the basic possibilities afforded by access to open source technical information will be illustrated using Iran's procurement of carbon fibre as an example.

Carbon fibre is of use in a huge number of applications, including in defence and aerospace, motorsports, and even fishing rods and tennis rackets. However, it is its possible use in centrifuges for enriching uranium and in missile applications that most often raises WMD-related concerns. Such applications typically require high-strength grades. The global manufacturing base for these materials is highly concentrated among the developed and industrialised economies.[16] This makes the material a key 'choke-point technology'. It is a material which it is both possible to prevent Iran from acquiring due to the narrow supply base, and without which the Iranian programme could encounter significant delays.

In August of 2011, Iranian state media reported that Iran had started manufacturing high-grade carbon fibre indigenously. Reports quoted Iranian Defence Minister General Ahmad Vahidi as saying that sanctions put in place by the international community prohibiting the export of carbon fibre to Iran had 'caused a bottleneck in Iran's production of advanced and smart defense systems', and due to this Iran was pursuing indigenous capabilities.[17] Along with published statements, Iranian state news outlet *Press TV* released a video including a sequence of shots of the carbon-fibre-production facility.[18] While providing valuable information, the source of the information obviously raises serious questions regarding its truthfulness.

When it comes to verifying technical information from open and closed sources, governments of larger states often possess technical resources. For example, the US has significant technical expertise at the National Labs, and other states may have expertise at state-owned manufacturing facilities or within their licensing systems.[19] However, in smaller, medium-sized and less well-resourced states, the available expertise is limited. In fact, across the board, by far the most extensive, relevant and up-to-date technical expertise can often be found within the private sector.

Continuing with the carbon-fibre example, there are a handful of technical consultancies that work to commission and optimise carbon-fibre production lines around the world. Consultation of these technical experts that possess tacit 'hands-on' knowledge of the operation of carbon-fibre-production facilities can be useful in determining the feasibility of Iran establishing an indigenous capability of a given quality, taking into account the restrictions on importing manufacturing equipment. This could be paired with information from academic and technical publications, which could provide insights into processes as well as Iranian research being conducted into carbon-fibre production and the open sources available to Iran's engineers, which could facilitate development of such a capability.[20]

Open source technical information is also of use to governments in assessing the risks posed by the production capabilities held by different firms around the world. When they are looking to market their products, firms in many industrial sectors provide datasheets regarding their capabilities. This open source information considered in conjunction with information gathered from industry experts, trade publications and other open sources can provide a relatively full picture of the manufacturing capabilities around the world, even without openly contacting firms to request further information. An assessment of global manufacturing capabilities derived from information from open sources can compliment more traditional intelligence held by states. This can allow national authorities to target their outreach to firms that are most likely to be targeted by proliferators.

It is clear that states, although having the largest resources from which to draw intelligence, possess an incomplete intelligence picture regarding illicit procurement. Open sources held by the private sector provide significant opportunities to national authorities, if properly exploited, to enrich the picture provided by more traditional forms of intelligence regarding illicit trade. However, while the private sector holds information of use to governments, similarly, government holds information of use to the private sector, though much of this is politically sensitive or classified. Open sources present an opportunity here, providing valuable information that is not subject to sensitivities.

Facilitating the first line of defence: Open sources and the private sector

The utility of open sources to governments is perhaps eclipsed by its utility to the private sector. While governments have access to a full spectrum of intelligence assets and sources, including more traditional 'secret' intelligence, the private sector does not. Open sources can be of great use here in helping private sector entities to assess business risks. This section will explore their utility in more detail, in terms of both understanding the risks associated with illicit trade more generally and conducting customer due diligence.

A holistic understanding of risk

Overall, the use of open source information can provide firms with a greater understanding of the risks faced by their sector. This understanding can be broader in nature, relating to how different industrial sectors are being targeted by proliferators, and to benefit the programmes of certain countries. However, the real benefit of open source to firms relates more to understanding specific risks associated with illicit trade, in terms of both the techniques used by proliferators and the entities involved. This information, if used effectively, can help firms to look beyond the licensing system in order to consider risk more holistically and to mitigate against it through compliance systems.

In order to be most effective, a compliance system needs to be specifically tailored to reflect each company's business. Open sources, used in conjunction with best-practice guidance, can help firms to understand how to do this. They can help firms to better understand their products and markets, and which non-controlled goods could be sought by proliferators. There is also useful information available regarding the risks posed by new markets.

Open source can also help firms to understand more specific risks within their industrial sector. For example, in the US and the UK, there are a number of realities of export licensing which governments have struggled to communicate. Take, for example, electronic technologies to China or nuclear technologies to India. In this respect, export licensing statistics can help firms to understand when obtaining a licence could be problematic. Some countries release annual licensing reports, whereas in other cases, they produce documents that detail prosecutions, which can highlight where there may be potential difficulties or concerns.[21] Open sources can act as a business facilitator, helping firms to be realistic when informing potential customers of the possibility of obtaining a licence or the time that it is likely to take.

Due diligence: Beyond a list-based approach

In recent years there has been a significant increase in focus on due diligence across the business community. Obviously this has been seen among exporters in preventing the transfer of their goods to WMD proliferators or terrorists. However, increased interest in the use of sanctions as coercive tools in countering proliferation, and the prevention of illicit activities such as money laundering and terrorist and proliferation financing within the financial and service industries, has seen an increase in due diligence practices to avoid involvement in prohibited transactions.

In the export community, and elsewhere, much of the work relating to entities has centred on the use of lists. These are created and maintained at the national level – for example, the US's 'Entity list' and list of 'Specially Designated Nationals', and the UK's 'Iran list'. They are also created and maintained at the international level – for example, lists are often included in the appendices of UN sanctions resolutions and in EU sanctions regulations. These can be integrated into business processes manually or through automated screening software.

However, there are significant difficulties associated with a reliance on lists. Lists do not contain all entities that may pose a risk; in fact, they may not be updated very often, and they will not include entities that may be the targets of ongoing covert intelligence efforts of governments. Entities that are listed frequently change their names to avoid detection by companies. Take, for example, notorious Chinese proliferator Li Fang Wei who has been sanctioned by the US numerous times. A 2009 indictment listed 13 company names, 8 aliases, and 5 official looking letterheads that he had been using, presumably to avoid detection by screening systems.[22]

Open sources can clearly be of use to firms in conducting due diligence that goes beyond checking entities against lists.[23] Open sources can be of use in verifying bona fides of 'positives', which automated screening systems flag as potential matches, and as part of a substitute for automated systems altogether. One major issue with automated screening is that systems will undoubtedly raise concerns regarding many entities. Most of these will constitute 'false positives' or entities that share characteristics with listed entities yet are not those actually listed. While screening systems typically have a low false positive rate of 0–3 per cent, a large number of transactions can mean many false positives to clear.

Smaller firms with limited resources or limited exports may opt to conduct screening and due diligence manually. Here, firms can use open source techniques. A starting point is usually the company website, which can be used to check that the requested product is consistent with the company's product line. Firms frequently use Internet search engines, such as Google, to check whether entities are listed or have been mentioned in press reports.[24] The entity's digital presence can provide a host of opportunities to verify the entity's bona fides and identify inconsistencies. Commercial directories, trade-show listings, professional networking sites such as Linked In, and trade publications all provide such opportunities. Google Earth and Street View could, in some cases, even be used to verify that an address given for a business is as stated.

One difficulty in this regard relates to firms' ability in conducting such research in different languages. For example, a Mandarin linguist conducted an experiment in mid-2013, searching for a number of Li Feng Wei's listed companies.[25] From a number of Mandarin language websites and an open-access Chinese customs database, the linguist was able to generate a large number of leads, such as mobile phone numbers, email addresses and personal details, including social security numbers. The search identified and potentially implicated two non-listed individuals, including Li's deceased mother, and several unlisted companies closely linked to Li.

Available open source: Incomplete and difficult to verify

In exploring some of the many possibilities afforded by open sources in preventing illicit trade, it is clear that there are significant difficulties relating both to the scope of information available and in practically deriving these benefits. As in many areas where there is scope for the use of open sources, the picture that can be drawn with the available open source information is an incomplete one, presenting significant challenges. For example, relevant information may not be available, private sector entities may not be willing to share information for a number of reasons, or they may not be able to identify all suspicious enquiries without national authority assistance. Also, there are significant issues surrounding time sensitivities. The most valuable information for the private sector relating to entities, for example, may not be available in open sources until it is no longer of use.

The verification of information, by the national authority and by compliance officers in industry, can be difficult. Take, for example, 'suspicious enquiries' that would provide significant amounts of information regarding illicit procurement activities. What firms and governments deem to be 'suspicious' is highly subjective. In some cases there are obvious ways to verify bona fides. However, this is not always the case. Take, for example, the experiences of one UK electronics exporter, which highlights that while some enquiries may have suspicious characteristics, the assistance of the national authority is sometimes necessary to initially identify illicit trade.[26] In fact, in some senses, receiving a number of enquiries from distributors in response to a tender raised by an illegitimate Iranian end user via proxies may look quite similar to that raised by a legitimate end user.[27]

Operationalising open source methods

Besides determining the value of the open source information, there are also difficulties associated with putting open source processes into practice. A primary issue relates to information sensitivities. While open source information is non-classified and of relatively low sensitivity, there are still issues that need to be overcome. Take, for example, information regarding suspicious enquiries. Firms are often reluctant to associate themselves with illicit activities, even if the transactions go unfulfilled. National authorities often lack the opportunities to build trusting relationships with the private sector required to facilitate information-sharing. On top of this there is often a lack of awareness of the value that open source could have among firms.

Beyond sensitivities it is also an issue of a lack of time and resources on both sides. The use of open sources has often been cited as cheap compared with more traditional forms of intelligence, and in some senses this also goes for its use in countering illicit trade.[28] As an industry representative has noted, 'to share procurement data will not cost industry or the respective member state authorities money; to forward an email is as simple as a mouse click'.[29] However, in a time of austerity and government cutbacks, there is limited appetite to pursue open source gathering efforts that may lead to new costs.[30] Similarly, private sector actors can be reluctant to bear the costs of a beyond-compliance approach. This is especially the case when there are few resources and guidance that allow them to pursue it in a best-practice manner.

Present and future uses of open source in countering illicit trade

Given both the clear utility of open-source information in countering illicit trade and the difficulties listed above, this section will consider cases where open source processes have been used, the limitations from these uses and what lessons may be learned. It will address the current uses of open sources by two international organisations, before moving on to consider some new approaches to using open sources in order to better counter illicit trade.

Open source in practice: UN expert panels

In some cases when UN sanctions resolutions are passed by the Security Council, a Panel of Experts is set up to monitor and report on implementation of the sanctions measures. There are three panels concerned with illicit trade issues, working to scrutinise the implementation of UN Security Council resolution 1540 (UNSCR1540) – a resolution that sets out a framework to prevent the involvement of non-state actors in the acquisition of WMD – and sanctions resolutions on Iran (UNSCR1929) and North Korea (UNSCR1874).[31] These panels consist of between half a dozen and a dozen 'experts' from the P5 and other interested states who engage member government and industry stakeholders, examining provided evidence and interdictions when they take place.

Both UNSCR1929 and UNSCR1874 set out the panel mandate to 'gather, examine and analyse information from states, relevant United Nations bodies and other interested parties' regarding implementation.[32] Their work, most notably in the country-specific panels, feeds into a final report that is delivered at the end of the panel's mandate and includes recommendations.

In some senses, this final report is a product of gathered open source information and related analysis. While some governments do contact the panel with non-classified but sensitive information, a large quantity of the information used to prepare the report can be considered open source. The North Korea Panel reports note that beyond panel member's first-hand observations, the panel 'relies on' two types of information: 'information supplied by states and/or international organisations, officials, journalists and private individuals; and information found in the public domain'.[33] Significant information is gained from consultations with industry, with the 2010 Iran Panel report detailing that consultations were held with 4 private sector entities, and the 2012 report listing 14.[34]

The panel reports provide some insights into the open-source processes used to develop the report in a 'methodology' section. Both country-specific panels emphasise the importance of 'high evidentiary and methodological standards'.[35] The 2012 Iran Panel report notes that 'the Panel endeavoured to ensure that its findings were substantiated, and information contained in its reports derived from credible sources'.[36] The North Korea Panel reports note that 'in weighing the reliability of information, the Panel always keeps in mind the identity and role of the sources providing it'.[37] Verification of information is obviously a difficult yet required dimension of an open source process, particularly when dealing with sensitive topics such as illicit trade, and the activities of secretive states such as North Korea.

Part of the work of the Expert Panels is to raise awareness or 'promote and facilitate the implementation' of the sanctions measures.[38] In this regard they are not merely reporting on non-compliance; they constitute an actor in themselves, engaging on these issues, providing information and raising awareness around the world. Internally in the UN system, the mandate of the Iran Panel and North Korea Panel includes 'making recommendations' as to how the Security Council and states could better implement the provisions of UNSCRs. Panel members also partake in civil society and industry events to better inform stakeholders, and to encourage involvement in the panel's activities by providing information.[39] The 2013 report of the Iran Panel highlights 'outreach' as a priority.[40]

However, a significant limitation of the work of the UN panels is found in associated political sensitivities. This has clearly been demonstrated when final reports have been blocked, and less clearly where the content of final reports may have been 'watered down'. Obviously, while panel members are technically UN staff, they are often seconded or former national diplomats, and they may retain some national biases. In the past, publication of reports has been blocked entirely. The 2010 and 2011 North Korea Panel reports were never officially released and technically remain internal Security Council documents.[41] The 2012 report contains a footnote clarifying that the Chinese panel member 'would like to disassociate himself from the 2011 report', which was leaked in May 2011.[42] Similarly, Russian pressure allegedly blocked the release of the 2011 Iran Panel report.[43] Such pressure may also compromise the extent and completeness of released reports. There have been suggestions that concerns were raised by other Security Council members regarding the 2012 North Korea report, which was publically released.[44]

Open source in practice: IAEA illicit trade monitoring

The IAEA also has an ongoing open source effort, which relates to the gathering and exploitation of illicit trade data. However, this effort is largely maintained in support of safeguards. As discussed in Chapter 4, information regarding states' procurement of sensitive items can feed usefully into the safeguards process.

While states' legal obligations to declare imports and exports under safeguards largely relate to fissile material, there have been several efforts to broaden information to other sensitive goods. In 1993 the IAEA Board of Governors endorsed a Voluntary Reporting Scheme on imports and exports of non-nuclear materials.[45] The AP (2005) also contains requirements for signatory states to report the export of certain sensitive non-nuclear materials and goods.[46] This provided information is supplemented by open source information, which is gathered and analysed by part of the IAEA State Factors Team, previously called the Trade and Technology Analysis Team (TTA), which was established within the Department of Safeguards in 2004.[47]

In support of safeguards, the team analyses trade data to better understand illicit trade and to increase the chances of detecting ongoing illicit procurement activities and possible covert breaches of safeguards. In creating the team, the IAEA was seeking to centralise all existing information regarding illicit trade and to create an 'institutional memory' regarding such issues.[48]

Central to the team's work is the Outreach Programme, which began in 2006, which involves data collection from relevant industries in different countries regarding suspicious enquiries.[49] Copies of relevant documentation are collected from companies with the view that useful information can be extracted, not only from the details of requested products and entities requesting them, but also from how documents are formatted.[50] Practically, a computer system, the Procurement Tracking System (PTS), has been used for the storage, management analysis, and visualisation of collected data.[51]

Difficulties and limitations

There are a number of limitations on the utility of open source processes as demonstrated by the UN panels and the IAEA Department of Safeguards, despite their valuable contributions. Largely these limitations relate to the objectives that these initiatives are primarily meant to achieve, the political limitations of operating out of larger international organisations, and the limited resources available.

The objectives of the two initiatives limit their utility in terms of considering illicit trade more broadly. The UN panels are mandated to look at breaches of UN sanctions. This is broader in scope than consideration of the technology-based UN sanctions. For example, the UN North Korea Panel must spend time considering breaches of prohibitions on the export of 'luxury goods' and the Iran panel must consider the sanctions against Iranian shipping. The IAEA's work is primarily focused on the nuclear area, not considering missiles, chemical and biological weapons, or military goods. Within the nuclear area, attention is primarily paid to 'safeguards relevant technology', reflecting the Department of Safeguards' mandate.[52]

There are also clear political limitations imposed by working through a large intergovernmental body. This goes beyond that of presenting findings as seen with the UN panel example above. For example, there are limits on how information can be gathered. The IAEA Outreach Programme must obtain consent from member-state governments before contacting companies within that country to collect information.[53] There has been some difficulty in obtaining the consent of the US and other states.[54] These efforts have also been restricted by resource limitations. The efforts of both the UN panels and the IAEA are limited by finite human and financial resources.

Future uses of open sources?

Given the insufficiencies in the current uses of open source, it is important to consider how open source may better be operationalised. A number of government-affiliated and non-governmental actors have been conducting work in this area more broadly and have provided various views for the future of private sector engagement.[55] A number of key characteristics of any beneficial approach to the use of open source issues can be identified in terms of organisation, driving the process and resourcing it.

Organisation

Certainly, benefits have been highlighted in utilising a neutral 'third party' to facilitate information-sharing.[56] Certain aspects of an open source 'intelligence cycle' could benefit from being carried out by a third party, including collection and dissemination of intelligence products.[57] In terms of collection, use of a third party can help to overcome industry concern surrounding possible prosecution, and commercial sensitivities. Information can be adequately sanitised by the third party before being shared with government to aid broader intelligence efforts.

In terms of dissemination, a neutral third party could provide information which governments may not be able to. For example, a third party could highlight risks which governments are unable to due to political sensitivities, such as commenting on the adequacy of export controls or the nature of transhipment or diversion risks in other countries.

Central to these information processes are handling issues that are crucial in order to build 'trust' with the private sector and collect valuable information.[58] All data collected from industry must be carefully handled and securely stored in line with relevant data-protection legislation. Similarly, there may also be benefits to limiting access to the open source products during the dissemination phase. Take, for example, if a list of suspect entities were maintained. To avoid entities from finding their name on the list and taking steps to change it, a firewall would be necessary with access details being provided to a select group of vetted firms and individuals in the industrial sectors which manufacture sensitive technologies.

Driving the process

One difficulty that would need to be overcome relates to what would drive or motivate private sector actors to be interested in providing information and engaging with such a third party. Reputational risk would likely be an important factor in motivating firms to provide information by incentivising them with access to information to mitigate risks more holistically. Supply chain linkages could be useful here, with firms wanting to fully mitigate risks, ensuring that their business partners take similar measures to remain compliant and to share relevant information. This has been suggested in a recently published NSG good-practice corporate guideline document.[59]

Social corporate responsibility can also form a significant part of this, with many firms now wishing to take active measures to improve their image and practices. A policy in this area could extend to implementing best practice in sharing information – for example, that relating to suspicious enquiries. This has also been suggested in the NSG guidance document.[60]

Resourcing

In terms of resourcing such an open source initiative, there is scope for both private sector and government involvement. Governments could clearly derive benefits from such efforts, and may be willing to finance them. Similarly, there is clear scope for resources to be drawn from the private sector. Many companies already pay for access to relevant information in the form of denied party-screening systems or guidance and consultancy. However, private sector resourcing of such efforts would be highly dependent on their perceptions regarding the benefits.

Conclusion: The role for open source

There is more open source information available on illicit trade than ever before. Despite some difficulties associated with this, there is certainly a role for open source processes to play in preventing proliferation procurement. The need for effective information-sharing mechanisms to be established between governments and private sectors is clear.

In terms of the broader roles that open source information can play, in most cases it can provide good context and background information. This has been seen in some senses with the use of technical open source information by governments, and in building compliance systems within industry. However, if effective mechanisms are set up, open sources can be highly relevant in a time-sensitive and operational sense, for both governments and the private sector. For governments and international organisations, gaining further information regarding suspicious enquiries can help them to build a better picture of illicit trade and complement more traditional intelligence sources. For the private sector, timely information about suspicious entities, the technologies that they are seeking for use in different programmes and the methods that they use can help compliance officers in industry to better identify suspicious enquiries.

The increased use of open sources, in the manner discussed, can help to transfer the role of intelligence in preventing WMD proliferation from constituting the 'last line of defence', in the sense of covert intelligence operations and interdictions, to the 'first line of defence' in assisting industry in preventing illicit procurement.[61] One notable industry practitioner has even gone as far as to argue that information itself constitutes the 'first line of defence'.[62] With the potential for open sources to play a role being so great, and the implications of proliferation being so serious, more effort needs to be invested in finding efficient processes and solutions to harness the role of open sources.

Notes

1. Christopher Hope, 'MI6 Chief Sir John Sawers: "We Foiled Iranian Nuclear Weapons Bid"', *The Telegraph*, 12 July 2012, http://www.telegraph.co.uk/news/uknews/terrorism-in-the-uk/9396360/MI6-chief-Sir-John-Sawers-We-foiled-Iranian-nuclear-weapons-bid.html.
2. On Iran through UNSCR1737 (2006) and on North Korea through UNSCR1695 (2006).
3. See, for example, the work of the Wisconsin Project, the Institute for Science and International Security, and Project Alpha at King's College London.
4. 'Frequently Asked Questions', Wikileaks website, http://wikileaks.org/static/html/faq.html.
5. Search conducted using Cablegatesearch, http://www.cablegatesearch.net/.
6. See, for example, the recent comments by Li Fang Wei: William Maclean and Ben Blanchard, 'Exclusive: Chinese Trader Accused of Busting Iran Missile Sanctions', *Reuters*, 1 March 2013, http://www.reuters.com/article/2013/03/01/us-china-iran-trader-idUSBRE9200BI20130301.
7. 'Frequently Asked Questions', Wikileaks website, http://wikileaks.org/static/html/faq.html.
8. Gordon Corera, *Shopping for Bombs* (NY, USA: Oxford University Press: 2006).
9. David Albright and Peter Gray, 'Building a Corporate Nonproliferation Ethic', *ISIS Report*, June 1993, http://isis-online.org/uploads/isis-reports/documents/ISIS_Report_Building_a_Corp_Nonprolif_Ethic_1.pdf.
10. Licence-application processes vary between different national export control systems but are largely based on similar principles and concepts.
11. UN, 'Panel of Experts Established Pursuant to Resolution 1929 (2010) Final Report', May 2011, p.22, http://a-pln.org/wordpress/wp-content/uploads/2011/05/2011-05UNpanelofexpertsreportonIran-1.pdf.
12. Ralph Wirtz, 'Industry Contribution to Thwart Illicit Nuclear Trade', remarks to the 2010 Safeguards Symposium, *IAEA-CN-184/198*, http://www.iaea.org/safeguards/Symposium/2010/Documents/PapersRepository/198.pdf.
13. David Albright, Paul Brannan and Andrea Scheel Stricker, 'Detecting and Disrupting Illicit Nuclear Trade After A.Q. Khan', *The Washington Quarterly* (2010), Vol. 33, No. 2, pp.85–106; p.102.
14. The SIEL is the most commonly issued licence type in the UK allowing the export of a quantity of stated goods to a single destination. Department for Business Innovation & Skills, 'Strategic Export Controls: Country Pivot Report, 1st January 2012–31st December 2012', https://www.exportcontroldb.bis.gov.uk/eng/fox/!STREAM?id=BU3Tox_E1nywMya6&stid=BU3Tog_E1nywMya6&app_mnem=sdb&mode=view&xfsessionid=sid_BU3Tof_E1nywMya6.
15. Ralph Wirtz, 'Industry Contribution to Thwart Illicit Nuclear Trade'.
16. See, for example, Ernst and Young, 'UK Composites Supply Chain Scoping Study – Key Findings', April 2010, http://www.compositesuk.co.uk/LinkClick.aspx?fileticket=Wm0WJ7RfFY0%3D&tabid=104&mid=532.
17. 'Iran Launches Production of Carbon Fiber', *Press TV*, 27 August 2011, http://www.presstv.com/detail/196011.html; Nasser Karimi, 'Iran Launches Production of Banned Carbon Fiber', *Associated Press*, 27 August 2011, http://news.yahoo.com/iran-launches-production-banned-carbon-fiber-070925257.html.
18. Video available from http://www.youtube.com/watch?v=tP_2HakdKCA, accessed March 2013.

98   OSINT and Proliferation

19. Technical expertise within the licensing system is likely to be confined to a small number of individuals with a wide remit to cover many technical areas.
20. Wyn Bowen, 'Open Source Intelligence and Nuclear Safeguards', in Robert Dover and Michael Goodman (eds.), *Spinning Intelligence* (London: Hurst and Company, 2009).
21. For prosecutions, see US Bureau of Industry and Security, 'Don't Let This Happen to You!', September 2010, http://www.bis.doc.gov/complianceandenforcement/dontletthishappentoyou_2010.pdf.
22. Supreme Court of the State of New York, 'Indictment: Grand Jury of the Court of New York vs. Li Fang Wei and Limmt Economic and Trade Company Ltd.', April 2009, http://graphics8.nytimes.com/packages/pdf/nyregion/08INDICT.pdf.
23. Jonathan Brewer, 'The Private Sector Plays an Important Role in Delaying the Development of the Iranian Nuclear Program', *The Bulletin of the Atomic Scientists-Online*, November 2010, http://www.thebulletin.org/web-edition/roundtables/iran-and-the-west-next-steps.
24. Daniel Salisbury and David Lowrie, 'Targeted: A Case Study in Iranian Illicit Missile Procurement', *The Bulletin of the Atomic Scientists* (2013), Vol. 69, No. 3, pp.23–30.
25. Experiment conducted by linguist at the Centre for Science and Security Studies at King's College London in June 2013.
26. Daniel Salisbury and David Lowrie, 'Targeted', pp.23–30.
27. Ibid.
28. Wyn Bowen, 'Open Source Intelligence: A Valuable National Security Resource', *Jane's Intelligence Review*, November 1999, pp.50–54.
29. Ralph Wirtz, 'Industry Contribution to Thwart Illicit Nuclear Trade'.
30. Jonathan Brewer, 'The Private Sector Plays an Important Role in Delaying the Development of the Iranian Nuclear Program'.
31. Officially described as the 'Security Council Committee established pursuant to resolution 1540 (2004)', the 'Panel of Experts established pursuant to resolution 1929 (2010)', hereafter referred to by the author as the 'Iran panel', and the 'Panel of Experts established pursuant to resolution 1874 (2009)', hereafter referred to the 'North Korea Panel'.
32. UN, 'Security Council Resolution 1929' (10 June 2010), paragraph 29; UN, 'Security Council Resolution 1874' (12 June 2009), paragraph 26.
33. Similar wording in UN, 'Report of the Panel of Experts established pursuant to resolution 1874 (2009)', 14 June 2012, p.10, http://www.fas.org/nuke/guide/dprk/unsc-june2012.pdf; UN, 'Report of the Panel of Experts established pursuant to resolution 1874 (2009)', May 2011, p.7, http://www.ncnk.org/resources/publications/UN-Panel-of-Experts-Report-May-2011.pdf.
34. UN, 'Report of the Panel of Experts established pursuant to resolution 1929 (2010)', May 2011, p.6, http://a-pln.org/wordpress/wp-content/uploads/2011/05/2011-05UNpanelofexpertsreportonIran-1.pdf; UN, 'Report of the Panel of Experts established pursuant to resolution 1929 (2010)', 12 June 2012, pp.9–10, http://www.securitycouncilreport.org/atf/cf/%7B65BFCF9B-6D27-4E9C-8CD3-CF6E4FF96FF9%7D/s_2012_395.pdf.
35. UN, 'Report of the Panel of Experts established pursuant to resolution 1929 (2010)', 12 June 2012, p.6; UN, 'Report of the Panel of Experts established pursuant to resolution 1874 (2009)', 14 June 2012, p.10.
36. UN, 'Report of the Panel of Experts established pursuant to resolution 1929 (2010)', 12 June 2012, p.6.

37. UN, 'Report of the Panel of Experts established pursuant to resolution 1874 (2009)', 14 June 2012, p.10.
38. 'Panel of Experts established pursuant to resolution 1874', UN website, http://www.un.org/sc/committees/1718/panelofexperts.shtml.
39. Panel reports include accounts of different organisations engaged. Panel members also partake in industry conferences and events.
40. UN, 'Final Report of the Panel established pursuant to resolution 1929', 5 June 2013, p.9, http://www.securitycouncilreport.org/atf/cf/%7B65BFCF9B-6D27-4E9C-8CD3-CF6E4FF96FF9%7D/s_2013_331.pdf.
41. Justin McCurry, 'North Korea "Is Exporting Nuclear Technology"', *The Guardian*, 28 May 2010, http://www.theguardian.com/world/2010/may/28/north-korea-exporting-nuclear-technology.
42. UN, 'Report of the Panel of Experts Established Pursuant to Resolution 1874 (2009)', 14 June 2012, p.11; Tania Branigan, 'China Denies Role in North Korea-Iran Missile Trade', *The Guardian*,18 May 2011, http://www.theguardian.com/world/2011/may/18/china-denies-role-north-korea-iran-missile-trade.
43. George A. Lopez, 'Russia and China: Sabotaging U.N. with Vetos', CNN.com, 8 February 2012, http://edition.cnn.com/2012/02/08/opinion/lopez-russia-sanctions-cold-war/index.html.
44. Mark Hibbs, 'China and the POE DPRK Report', Armscontrolwonk.com, 2 July 2012, http://hibbs.armscontrolwonk.com/archive/879/china-and-the-poe-dprk-report.
45. Richard Hooper, 'The System of Strengthened Safeguards', *IAEA Bulletin* (1997), http://www.iaea.org/Publications/Magazines/Bulletin/Bull394/hooper.html.
46. C. Vesino et al., 'Global Trade Date to Support IAEA Safeguards', *ESARDA Bulletin* (2010), No. 45, http://esarda2.jrc.it/bulletin/bulletin_45/B_2010_045_05.pdf.
47. Matti Tarvainen, 'Unfair Trade', *IAEA Bulletin* (2009), Vol. 50, No. 2, http://www.iaea.org/Publications/Magazines/Bulletin/Bull502/50203556163.html.
48. Matti Tarvainen, 'Procurement Outreach in Revealing Proliferation Networks', Presentation to the Carnegie Endowment for International Peace, June 2007, http://carnegieendowment.org/files/illicit_networks_tarvainen.pdf.
49. Matti Tarvainen, 'Nuclear Trade Analysis May Provide Early Indications of Proliferation', *ESARDA Bulletin* (2008), No. 4, http://esarda2.jrc.it/db_proceeding/mfile/B_2008_040_09.pdf.
50. Matti Tarvainen, 'Procurement Outreach in Revealing Proliferation Networks'.
51. A. El Gebaly et al., 'Analysis and Processing Tools for Nuclear Trade Related Data', in the IAEA's *Proceedings of an International Safeguards Symposium on Addressing Verification Challenges*, 16–20 October 2006, http://www-pub.iaea.org/MTCD/publications/PDF/P1298/P1298_Contributed_Papers.pdf.
52. IAEA, *Safeguards: Staying Ahead of the Game* (Vienna, Austria: IAEA, 2007), p.20, http://www.iaea.org/Publications/Booklets/Safeguards3/safeguards0707.pdf.
53. Mark Fitzpatrick, 'Understanding Clandestine Nuclear Procurement Networks', presentation to the IAEA Scientific Forum, 19 September 2007, http://www-pub.iaea.org/mtcd/meetings/pdfplus/2007/cn159/cn159_fitzpatrick2.pdf.
54. David Albright, Paul Brannan and Andrea Scheel Stricker, 'Detecting and Disrupting Illicit Nuclear Trade After A.Q. Khan', pp.85–106; p.101.
55. See, for example, the work of Project Alpha at King's College London (UK), Pacific North-Western National Laboratory and the Institute for Science and International Security (both US-based).

56. Ian Stewart, 'Antiproliferation: Tackling Proliferation by Engaging the Private Sector', Belfer Centre Discussion Paper, November 2012, http://belfercenter.hks.harvard.edu/publication/22460/antiproliferation.html.
57. OSINT intelligence cycle concept taken from NATO, *NATO Open Source Intelligence Handbook* (NATO, 2001), http://www.oss.net/dynamaster/file_archive/030201/ca5fb66734f540fbb4f8f6ef759b258c/NATO%20OSINT%20Handbook%20v1.2%20-%20Jan%202002.pdf.
58. Andreas Widl, 'Non-proliferation: Social Responsibility in Industry', *1540 Compass* (2012), Vol. 1, No. 1, pp.32–34.
59. Nuclear Suppliers Group, 'Good Practices for Corporate Standards to Support the Efforts of the International Community in the Nonproliferation of Weapons of Mass Destruction', NSG Website, http://www.nuclearsuppliersgroup.org/A_test/01-eng/NSG%20Measures%20for%20industry%20update%20revised%20v3.0.pdf.
60. Nuclear Suppliers Group, 'Good Practices for Corporate Standards to Support the Efforts of the International Community in the Nonproliferation of Weapons of Mass Destruction'.
61. David Albright, Paul Brannan, and Andrea Scheel Stricker, Detecting and Disrupting Illicit Nuclear Trade After A.Q. Khan', pp.85–106; p.94.
62. Andreas Widl, 'Non-proliferation', pp.32–34.

Part III

Open Source Intelligence and Humanitarian Crises

6
Positive and Negative Noise in Humanitarian Action: The Open Source Intelligence Dimension

Randolph Kent

> Whenever a Scud missile was launched from Damascus aimed at Aleppo, we could hear the sound coming from the army base nearby, and we would put this on Facebook and say that a missile was launched at such and such a time, and then someone nearer Aleppo would pick this up from Facebook, and then also put this information on Facebook. And this is the way that we warned each other. We also used Facebook to let people know where there were medicines and other relief supplies.[1]
>
> Syrian refugee, 2012

Introduction

Social networking has for the most part been greeted as a transformational breakthrough in the world of humanitarian action. Faced with a growing number of humanitarian threats and their ever more complex dimensions and dynamics, those with humanitarian roles and responsibilities increasingly rely on the Internet and its multiple by-products to deal not only with humanitarian crisis response but also with prevention and preparedness. Ushahidi, Twitter, Facebook, open source software and radio-in-a-box technology are just some of the myriad innovations that are transforming what since the early days of the Cold War has been called open source intelligence. In the humanitarian sector, these new forms of open source communications provide hitherto unimagined opportunities to anticipate actual and potential crises, to determine needs and the impact of assistance, and to identify and support creative measures for a range of humanitarian activities.

Yet, while the potential impact of this expanded form of OSINT is positive, there is also a negative side. There are an increasing number of incidents where the information gleaned from social networking has added confusion to humanitarian operations, where inconsistencies and contradictions have made it difficult for those seeking to assist to determine priorities when it comes to identifying beneficiaries and their needs. It is very likely that this sort of 'negative noise' will intensify as mobile technologies become ever more pervasive, and as more and more humanitarian threats and crises compete for space.

With this in mind, this chapter will explore four distinct, but overlapping, dimensions of OSINT as a factor in humanitarian action. The first relates to what may be described as the information 'baseline' that had until the past decade determined the sources of information that humanitarian organisations used to anticipate and respond to humanitarian crises. That information 'baseline' had, however, been built upon a set of Western hegemonic assumptions that had begun to erode towards the end of the twentieth century. Consequently, the second theme concerns how that erosion was in part driven by and drove a new information age for the humanitarian sector. The chapter's third theme looks at that new information age or, more specifically, at the impact that a range of new information technologies that circumscribed new forms of OSINT had upon humanitarian activities. As mentioned, the new technologies that underpinned a growing amount of OSINT were impressive, indeed transformational, but not without their downsides for those with humanitarian roles and responsibilities. Over the past few years, these have been increasingly recognised; some solutions have been found, but many challenges remain. In this context, the fourth theme focuses on some painful lessons learned and some more to come.

In the beginning: The information baseline

Since its nineteenth-century origins, modern humanitarianism has been underpinned by moral rectitude and economic dominance. Western-centric and shaped by the experiences of industrialism, colonialism and the notion of a civilising mission, among other things, the modern humanitarian sector proclaimed the universality of its principles, and, more subtly, the presumption that it understood the vulnerabilities as well as the requirements of 'disaster victims' and 'beneficiaries' better than they did themselves.[2]

Economic dominance, in particular, reinforced the divide between those perceived as hapless victims and those deemed to be resilient in various ways. Moreover, the moral rectitude driving humanitarianism was nourished by the notion that economic stability and superiority were products of a strength and resilience that those needing aid did not possess. There was a sense of duty towards those in need of aid. At the same time, however, this divide also reflected deeply rooted power relations. Those providing aid presumed that they knew best what so-called 'beneficiaries' needed, and these presumptions determined the sorts of information about humanitarian needs that, in turn, determined humanitarian responses. Indeed, the control of information about the requirements of crisis-affected peoples – both intentionally and unintentionally – was fundamental to the humanitarian sector. Such control not only reflected the dynamic of power between those providing assistance and those receiving it, but also reflected competing interests within the humanitarian structure itself. Information in both instances was, and remains, power.

In this context, it was perhaps inevitable that as humanitarianism gained momentum in the second half of the twentieth century and the number of organisations providing and coordinating aid multiplied, power relations within the humanitarian structure contributed to the emergence of a powerful form of economic rationalism that saw more attention given to the source and supply of aid than the victim.

The response to the 2004 Asian tsunami provides a recent example of this trend. In reviewing the international community's response to the tsunami, in which an estimated 228,000 people were killed, a multi-agency evaluation noted that 'those who own a process control it; they decide which priorities and policies apply'.[3] This was certainly the case in the ensuing relief operation in 2004–2005.[4] As the 2007 Tsunami Evaluation Coalition (TEC) report reminded its readers, affected populations cannot exercise effective ownership when international responders are not transparent with information and do not hold themselves accountable to the affected population.[5]

While in this instance it was noted that an unusually large number of surveys of affected populations were undertaken, those surveys were rarely translated into action.[6] Arguably of more importance to those responsible for the relief operations was the information provided to donors; 'accountability' was less focused on the tsunami-affected and more upon the sources of assistance. Herein lies the conundrum for those in need, those institutions that seek to provide assistance and those organisations that support such assistance: it is deemed essential by providers as well as supporters to control information in order to ensure the continuation of an adequate flow of resources, as well as the system through which they are provided. It has been noted, for example, that international NGOs have tended more and more to side with the donors upon whom they depend, and have become more distant from 'the poor beneficiaries they sought to assist'.[7]

Hence information in so many ways reflects the preferences and approaches of donors and not necessarily those of the affected. Though not protected as traditional intelligence might be, the ways in which information is conceived, guarded and presented reflects what might be termed a kind of 'information protectionism'. In a similar vein, a supply-driven response to humanitarian crises often combines a 'we know what is best for you' attitude with what donors are willing to offer or to fund. Furthermore, this approach permeates the humanitarian system. Look no further than the long-running and now well-documented competition between UN agencies and NGOs, for example.

In the words of the humanitarian director of a major NGO, 'The humanitarian ethos has been lost to an "economic rationalist agenda"'.[8] Of course, it would be reductive to extend this statement to the entire humanitarian sector; there are undoubtedly exceptions. However, there is certainly a sense among observers as well as many within the sector that humanitarian commitment over the past four decades has increasingly been weighed against the need to ensure the perpetuation of the organisation. Humanitarian intervention is a lucrative industry and commercial interests reinforce the idea that organisational survival is essential to ensure the wellbeing of others.

In this environment, information flows play an important role. Traditionally, sharing information across humanitarian actors has been notoriously problematic. This is due, in part, to the myriad ways in which information about the sources of a disaster, its impacts and its effects are gathered. There is also the difficulty associated with identifying relevant information across different specialisations and disciplines. Finally, and perhaps most important, information is protected from other organisations because its value can be calculated in terms of raising funds, demonstrating comparative advantages and, ultimately, consolidating the power and position of an organisation.

It is therefore gratifying that UN Consolidated Appeals – intended to inform potential donors about emergency relief requirements – have recently begun to demonstrate links between individual agency's operational activities. Similarly, efforts to promote coherence and cooperation resulted in a system developed by former UN Emergency Coordinator Jan Egeland in 2005 comprising clusters – groupings according to specialisation. This system was devised to overcome some of the more obvious difficulties in sharing information. Even in assemblages of specialists such as the clusters, information is too often protected by humanitarian agencies from humanitarian agencies.

Changing contexts, new technologies and OSINT

If Western hegemonic influence dominated much of humanitarian action throughout the late twentieth century, the adjustment, if not the decline, in that influence continues to rebound on humanitarianism in a variety of ways. Humanitarian assumptions about the universality of its principles, rights to access crisis-affected areas whether in conflict or not, 'the hapless south and the resilient north', the capacity to identify, prioritise and deliver life-saving assistance, have all, in one way or another, been challenged. One such challenge has inadvertently stemmed from the emergence of Internet-based technologies, which have altered not only the ways in which humanitarian assistance is identified, but also the ways in which it is delivered. The changing context in which humanitarian action operates and the impact of new technologies are intricately intertwined with the intelligence derived from information that is open to all and, in a sense, comes from all – OSINT.

The changing global context

The changing global context proceeds from the assumption that global change is not happening incrementally but rather exponentially. For the humanitarian sector, this has to be seen not only in terms of the decline of Western hegemony and the rise of new powers such as China, but also in the context of (a) the political centrality of humanitarian crises; (b) the globalisation paradox; (c) the resurgence of sovereignty; and (d) emerging technologies and their consequences:

Political centrality of humanitarian crises. Today, humanitarian crises have far greater political significance than they had in much of the latter part of the twentieth century. As Hurricane Katrina in 2005 and the Deepwater Horizon oil spill five years later demonstrated, even the most powerful governments have to deal with serious reputational issues if they fail to respond adequately to humanitarian crises. As humanitarian crises move to centrestage of governmental interests, they are imbued with high levels of political significance – both domestically and internationally. Indeed, it is no exaggeration to say that political survival, certainly at the ministerial level, may depend on the nature of the governmental response. Furthermore, the fallout is not limited to the government of the affected country. Increasingly, the ways in which neighbouring governments and other international actors respond to humanitarian disasters have important political consequences. That scrutiny – whether it stems from China's 2008 Sichuan earthquake or the 2013 Hurricane Sandy on the US East Coast – is intensified and, in many instances, driven by growing public ability to access and provide information across a range of outlets.

The globalisation paradox. This dictates that the more globalised the world becomes, the more 'localised' it will also be. This is increasingly reflected in new waves of 'state-centrism', in which the growth of global commonalities and inter-relationships have provoked an often intense reaction on the part of nations determined to protect their customs, culture and language. In this regard, open information flows reinforce the growing ability of local communities to express their views about the consequences of specific hazard impacts and the solutions required to address them from their own cultural perspectives and in their own languages.

Governments of crisis-affected states are therefore becoming increasingly wary of those outside humanitarian organisations that feel that their biggest contributions will stem from 'boots on the ground'. In those instances where external involvement is acceptable, prerequisites might include proven competencies in local languages and an appreciation of local culture. Increasingly, external assistance will be driven less by supply and more by demand, and the conduit for such assistance might well be through acceptable regional organisations or the private sector rather than the UN system or Western consortia.

Resurgence of sovereignty. Localisation in this context at the same time reflects a growing resurgence of sovereignty. Who interprets what is needed for a humanitarian response and how this need will be provided for will be one clear demonstration of sovereignty. Governments will be more inclined to resist unwelcome though well-intentioned external intervention, and will also be more insistent on determining whether or not external assistance is required and, if so, what will be provided, by whom, when, where and how.

For traditional humanitarian actors, the consequences of more assertive sovereignty mean that there will be even less receptivity to arguments about rights of access, that alternative providers (i.e. non-traditional actors, including the private sector) might be preferred 'humanitarians.' In various ways, governments' control over, for example, the private sector would be more feasible than that over the relatively free-wheeling autonomous humanitarian agencies, such as international NGOs.

Emerging technologies and their consequences. The hazards that emerging technologies create as well as their positive impacts are well recognised. Nevertheless, their longer-term consequences present profound unknowns. Unmanned aerial vehicles, including 'drones', cybernetics and space, nanotechnology, artificial intelligence and the much-vaunted 'social networking' phenomena present a vision of possibilities that are profoundly transformative, and yet their social, socioeconomic and political consequences are redolent with uncertainty. For humanitarian organisations, the interaction between an ever-increasing range of technologies and natural hazards will pose ever more challenging strategic and operational issues. And in this context the implications of the technologies that have accelerated open source intelligence and its impact have been transformational, though their full implications remain profoundly uncertain.

Emerging communications technologies in community contexts

At the beginning of the twenty-first century, rinderpest, a viral disease affecting cattle, occurred in many parts of Somalia, adding even more economic and social challenges to that beleaguered country. The Saudi Arabian market, so vital for Somali traders, was to be closed off until a solution to the cattle disease could be found. Few immediate solutions were apparent, in no small part because of the sheer difficulties of operating in what had been deemed to be an insecure 'failed state'. Under such circumstances, standard institutional procedures – including those of the UN's Food and Agriculture Organisation – could not respond with sufficient speed and coverage. Yet, to the surprise of international officialdom, a network of veterinarians from one end of strife-ridden Somalia to the other began to monitor the epidemic and provide solutions – via their mobile phones. Somalia had, in the midst of so much chaos, without any federal communications authorities, demonstrated the impact of emerging communications technologies.

From crowd-sourcing to social network analytics, the range of such technologies and their application is too numerous to attempt to list here. It is worth noting, however, how several such technologies have already begun to transform people's lives and livelihoods around the world. Their common denominator is that all reflect information that is, at least in principle, universally accessible. Each in its own way reflects technologies that change the ways in which people live, potentially 'empowering', and each in its own way has humanitarian consequences – for, if up to now Western-hegemonic institutions acted as filters for information about crisis impacts and needs, that role is undergoing radical change due to technologies that are transforming not only access to information but also accountability.

By 2025, the majority of the world's population will, in one generation, have gone from having no access to unfiltered information to accessing all of the world's information through a device that fits in the palm of the hand... On the world stage, the most significant impact of the spread of communication technologies will be the way they help reallocate the concentration of power away from states and institutions and transfer it to individuals.[9]

As the examples below suggest, the use of communication technologies are becoming norms, fundamental to the ways in which more and more people in communities pursue their lives and livelihoods in urban as well as rural areas. The interface between such technologies and humanitarian action is merely an extension of that norm and its impact upon societies as a whole.

Crowd-sourcing and crowd-funding. The impacts of crowd-sourcing and crowd-funding – as well-recognised means to promote causes and related resources online – have already been the subject of considerable analysis.[10] From the Arab Spring to business solutions, crowd-sourcing and crowd-funding have been seen as means of challenging the constraints of conventional institutions, and gathering opinions, expertise and resources more spontaneously and speedily than through standard organisational systems. In the world of development and humanitarian action, crowd-sourcing and crowd-funding have been interpreted as a 'positive threat' to the approaches used by UN agencies, bilateral donors and NGOs. 'They by-pass the middleman', noted one NGO representative responsible for fund-raising. 'People interested in a project can get together via the Internet, fund it and engage directly with those communities that wanted it. They increasingly won't need us.'[11]

In business, too, firms such as the Massachusetts-based open innovation company InnoCentive develop products by drawing on the creativity of hundreds of thousands of interested online followers to find solutions to complex problems. Crucially, these potential problem-solvers are not geographically limited; they are distributed around the world. They are not assessed in terms of degrees or social status but solely according to their individual abilities 'to get online'.

Enhanced means of livelihood. According to one Ethiopian farmer, he can negotiate the price of his *teff* grain through his mobile phone. No longer is he dependent upon local merchants as sole outlets for his goods. He can now negotiate with various merchants outside his immediate *woreda*, or administrative region, and determine who will offer the best price for his crops. More integrated systems use mobile phones to interface with web-based commodities exchanges offering product and price information directly to producers. An example is the Livestock Information Knowledge System (LINKS), which provides price and sales volume data for cattle as well as climate information in Kenya, Ethiopia and Tanzania.[12]

In another example, a handful of librarians recently helped 100,000 farmers obtain some USD187 million in subsidies via new Internet and computer services in a period of just two years, according to IREX.[13] The Biblionet-affiliated librarians, peppered throughout 30 counties in Romania, helped farmers to learn how to use the technology to access financial forms and submit them to the government. These efforts have enhanced productivity without burdening farmers with the costs of lengthy trips to administrative centres and related expenses.

The Colombian town of Aguablanca was marked by high levels of unemployment, high health risks, low levels of formal education, and violence. Not only would potential employers do little to invest in the area, but the stigma of being part of the poorer sections of the town meant that even those looking for employment were discriminated against because of where they came from. A major effort to provide access to the Internet via telecentres gave a dedicated number of those same people the opportunity to try to change potential employers' perceptions about the town's economic potential. While these efforts had only limited success, the importance of the telecentre initiative was that it gave a voice to what hitherto had been a silent community.[14]

Resource transfers and transformations. Increasingly the Internet and mobile technology are transforming assumptions about the availability and types of resource open to those who had been regarded as the dispossessed. For example, Equity Bank's engagement in cash transfer programmes across East Africa sits within its core business of providing payment solutions to Kenya's unbanked and underbanked. And it is through mobile communications systems that Equity Bank has been able to develop a product that can help it both meet development and humanitarian needs and expand to new markets within the country.[15] In a related vein, mobile phones have become an integral part of cash-transfer schemes that enable the urban poor to purchase food, and in rural areas of Bangladesh the NGO known as BRAC can provide a very popular microcredit scheme because of the ubiquitous use of mobile technology.[16]

On another front, the emergence of diasporic communities as a driving factor in many countries' economies and social support has been fuelled in no small part by the remittances which are transferred via the Internet. Money sent home by migrants constitutes the second largest financial inflow to many developing countries, significantly exceeding development assistance. According to the World Bank, USD401 billion worth of remittances went to developing countries with overall global remittances (including developed countries) topping USD514 billion.[17] While in the recent past such transfers have been undertaken via physical transfer houses, more and more of them are being forwarded by the Internet, allowing emigrants to take direct action to support needy communities in their country of origin.[18]

Educational access. While access to education remains a major problem throughout the world, there are a growing number of examples where Internet access is transforming approaches to education. A recent report prepared by mobile industry body the GSM Association gave examples of how mobile technologies are transforming educational systems in more and more places around the world.[19] It will soon be commonplace that children who, for reasons of distance, lack of fees, or political insecurity, are missing out on schooling will be able to gain access to education via a mobile phone: 'Even for those children without access to data plans or the mobile web, basic services like text messages and IVR (interactive voice response, a form of voice recognition technology) can provide educational outlets.'[20]

Global information networks at the local level. There is a range of global networks, many of which are satellite based, that have historically been available to a relative minority in order to transfer information. Government institutions, specialists including defence and meteorological departments, those in the world of the private sector and special interest groups have all been part of global information network phenomena that were confined to informational purposes. Now, however, technology has opened up information to all.[21] Information flows such as Google Earth bind global cities together in networks, creating a global city web whose constituent cities, linking emerging global with traditional ones, are creating new socioeconomic patterns – a new form of circuitry, which is no longer confined to specialists (Table 6.1).[22]

An old mantra about humanitarian action suggests that disasters are not aberrant phenomena, separate from the societies in which they occur. Rather, they are reflections of the ways in which societies function – the ways that they are structured and allocate their resources. In this same sense, humanitarian action takes place in societies around the world in which

Table 6.1 Global information networks: Past, present and future[23]

	Past	Present	Future
Purpose	To inform	To share	To engage
Core activity	Gathering information	Tracking issues	Guiding action
Stakeholders	State actors	State and private actors	Multiple actors
Information content	Discrete data	Networks	Relationships
Treatment of information	Data confidentiality	Privileged information	Transparency
Software tools	In-house capabilities	Analytical tools	Open source platforms
Output	Specialist assessments	Multiple stakeholder perspectives	Crowd-sourced opinions

transformations in communications technology and usage have radically altered some of the most fundamental aspects of day-to-day existence. Their effects on humanitarian action are an extension of those transformations.

Humanitarian action and OSINT

OSINT has revolutionised humanitarian action in at least four ways: (a) prevention and preparedness; (b) needs assessments and humanitarian response; (c) resource mobilisation; and (d) accountability. Each relates to what earlier was discussed as the changing global context in which humanitarian action operates; for those changes explain in various ways the receptivity that emerging norms are increasingly having with those who have humanitarian roles and responsibilities. Each, too, demonstrates that humanitarian action increasingly reflects 'the voice' that the crisis-affected have come to assume in their normal lives.

Prevention and preparedness. In Kenya the Sustainable Agriculture Livelihoods Innovations (SALI) initiative engages with 2000 farmers, based upon a 'demand-driven model', that uses an SMS platform developed by the Kenya Met Office. This is an interactive system, fuelled by the MPesa system,[24] which allows farmers and those in the Kenya Meteorological Office to exchange information about weather patterns – both short term and long term. One of the benefits of this ready exchange of information is that, with the knowledge and relative stability that this brings, farmers are more willing to invest in high-value crops. More importantly in this context, it serves as a means to prevent and prepare for crises that can put communities at risk and destroy livelihoods. In the Kenya case, for example, the threat of crop-devastating, caterpillar-like armyworms can be monitored by technology-driven communications to and from the communities and central authorities.[25]

In a related vein, communities in Senegal have established a similar system, thanks to easy access to SMS systems. The vulnerability of many communities to flooding has been significantly offset by an open source system of communications that links the government's meteorological offices to these communities, and it is the capacity to communicate regularly and in real time that has been seen as means to ensure that the full impact of flooding is prevented.[26] Similar systems exist in a growing number of countries, including Malaysia, where satellite-linked sensors are placed in flood-prone rivers that enable local communities to ascertain potential threat levels and communicate directly with emergency departments.

When it comes to preparedness, OSINT has now made it possible to establish systems that will in times of crises ensure that information can be shared. Such capacities are also important to provide the parameters of crisis preparedness by determining possible threats and opportunities to offset them. This form of preparedness has taken many forms. Easy access to information about preparedness is in and of itself a significant change. Social media has resulted in bringing together the public and private sectors with civil society, resulting in sustained stakeholder training and collaboration.

The Kenya Red Cross and the World Bank, for example, bring together disaster-relief experts and software engineers to work on ways to have greater interactive collaboration in identifying potential crisis threats and solutions. New York City's Office of Emergency Management uses the social media platform Sahana to prepare for shelter needs to manage at least 500 shelters that can accommodate up to 80,000 people. In other words, people's knowledge is drawn into the preparedness planning processes – a transformational change in the normal top-down approach to crisis planning.[27]

Needs assessments and humanitarian response. Critical to the relief process is the capacity to determine the extent of damage caused by a humanitarian crisis driver and the consequent needs of affected populations. Increasingly this is less and less the sole domain of 'experts' and more and more in the hands of affected communities. In this sense, there are a growing number of platforms emerging that can not only identify needs but also help to identify changing patterns of needs and their impacts on relief requirements.

This is evident across a range of crises. The office of the United Nations High Commissioner for Refugees has, as of 2009, engaged with an extensive network of social media that enables it to have two-way dialogues with refugees. In almost every crisis, a feed is created which can generate discussion flows through Flickr, Twitter, Facebook and YouTube – all increasingly available to populations despite the fact that they are displaced. The Dutch government operates a dedicated website (http://www.crisis.nl) as a focal point for the public dissemination of information during a disaster. In the aftermath of the 2010 Haiti earthquake, the Ushahidi platform collected information from crowd-sourced data and mapped it to assist relief workers. Similarly, SensePlace2, another map-based web application, integrates multiple text sources, including news, RSS and blogs, to filter through information that helps to monitor changing needs and the progress of relief operations.[28]

In the same way that needs assessments have been transformed by the technology that underpins open source information, crisis response too has changed in very significant ways. The sense of haplessness and helplessness that all too often burdened relief recipients has to some extent been reduced by the empowering effects of cash credits in times of crises. MasterCard, for example, is working with the World Food Programme on credits for the purchase of food in designated shops, and, as was noted earlier, Equity Bank has similar schemes only for a wider range of goods.

Open source information can also be used to determine if and when relief items have arrived, and the remaining gaps that need to be filled. This sort of crowd-sourced information can in various instances give a fuller picture of the state of operations more quickly than conventional operational monitoring. Provision of health care is also benefitting from technologies that enable diagnoses to be made even when distances between patient and doctor are long and expertise is limited. Telemedicine, still in its infancy, holds out the prospect for individuals to receive medial assistance, including operations, through direct mobile communications with 'health providers'.

Resource mobilisation. The conventional humanitarian sector has dedicated considerable energy over the past four decades to raising resources, in part for relief efforts and in part, too, to ensure its own institutional survival. Here, however, is another area where OSINT and the technologies that sustain it have considerable potential to change the relationship between aid providers and recipients. For example, the disaster-affected and those with resources to assist may no longer require what earlier in this chapter was referred to as the intervention of 'the middle man' to act as a conduit for assistance. A US NGO representative in charge of fund-raising recently highlighted the potential implications of direct engagement that bypasses the intermediary when she noted that there 'will soon be a day when a crowd-sourced appeal from a crisis area will be heard by crowdfunders who will not see the need to go through us to provide relief'.[29]

Indeed, moves in this direction have already begun with communities of diaspora acting as direct contributors by sending funds through mobile networks to designated recipients, mainly families. Moreover, the 2010 Haiti crisis illustrates the growing importance of these alternative sources of emergency funds. Contributions sent directly to Haiti by the Haitian diaspora residing in the US became so important in the aftermath of the earthquake that the US government offered a temporary amnesty to Haitian citizens working in the US illegally. This was to ensure that the flow of remittances continued to the disaster-affected.[30]

The importance of the Internet in the context of resource mobilisation for humanitarian needs is now widely recognised. Indeed, for almost two decades now, from the 1998 floods in Grand Forks, North Dakota, and the ravaging Hurricane Floyd in 1999, to the 2001 terrorist attack on the US, and the destruction of Hurricane Isabel in 2003, there has been a consistent increase in the use of the Internet to raise funds to respond to disaster needs. This trend was acknowledged in 2005 by a US committee focused on establishing a successful relief fund in times of domestic disasters: 'you will miss a critical resource if you opt not to use the Internet'.[31]

This form of resource mobilisation continued in the aftermath of Hurricanes Katrina and Russ where, according to the Pew Foundation, 13 million Americans made donations to relief efforts online and 7 million set up their own hurricane relief efforts using the Internet.[32]

Humanitarian action and accountability. Accountability in the humanitarian context is not merely about the effects and impacts in the aftermath of relief operations. It is also about holding to account those with humanitarian roles and responsibilities for the actions that they could and should have taken to prevent or prepare for humanitarian crises. Traditionally, as noted earlier in this chapter, those who often have the best understanding about threats and means to offset them are the vulnerable themselves and, all too often, 'experts' do not engage with them. Or, in the words of one victim of Hurricane Katrina in 2005, 'It don't seem that experts like talkin' to the poor.'[33]

And yet, despite declarations of commitment to community engagement over the years, there remains a seeming paradox – namely, that the more professional humanitarian actors become, the less inclined they appear to be to engage with the vulnerable and crisis-affected. These populations are still too often treated as passive onlookers, as experts determine not only what they should do in times of crises but also what they require in the aftermath of crises. As noted in a recent analysis of innovation in the humanitarian sector, 'Currently, humanitarian organizations – responsible for implementing projects over a relatively short time frame [usually 12 to 18 months] – have little time to observe and reflect on the profile and changing needs of their "customers" and on the efficacy of their implementation of goods and services.'[34]

It is here that the impact of OSINT is also increasingly being felt. Utilising a suite of open source technologies, including hosted online forums, social media and geomapping, a group of humanitarian organisations are assisting a range of Somalis 'to foster beneficiary participation in development and humanitarian interventions, by encouraging beneficiaries to express their demands, aspirations, engaging in the process of formulation of humanitarian interventions, planning, monitoring, and evaluation'.[35] In that sense, improving accountability is no longer a choice; it is a necessity. Paul Knox-Clarke, head of research and communications at the UK-based Active Learning Network for Accountability and Performance (ALNAP), sums up the point: 'Now [that] disaster-affected populations can make their voices heard through the media, agencies are recognizing that forward accountability [as in, accountability to disaster-affected communities] is no longer an option.'[36]

Haitians, for example, used Twitter to voice their frustration about slow aid responses to the 2010 earthquake. A group of Haitians (with help from the US-based Institute for Justice and Democracy) is taking the UN to court for allegedly fuelling a mass cholera outbreak, which went on to infect 500,000 people and cause 7,000 deaths, according to Medecins Sans Frontieres (MSF).

Aid survivors and diaspora communities are directly and vocally influencing aid responses. Survivors of Indonesia's Mount Merapi eruption in 2010 used Twitter, SMS and Facebook to tell the world what was happening, and sent their key findings to the UN Office for the Coordination of Humanitarian Affairs. The Somali diaspora responded directly to funding needs for food aid posted on an interactive map by the Danish Refugee Council. Meanwhile, hundreds in the Haitian diaspora collaborated to create a map of hardest-hit areas using an online tool called OpenStreetMap.[37] In each of these instances, the voices of communities are changing the dynamics of accountability at the front end as well as in the aftermath of humanitarian crises.

Whether it be transformations in the ways that the affected can hold providers to greater account, or the ways that prevention, preparedness, response and even resource mobilisation are beginning to be transformed, there can be little doubt that at the heart of so many of these fundamental changes are the methodologies that underpin OSINT.

The downside of up: 'Noise', source control and mixed messages

At the time of writing, a major debate is emerging about the perceived abuse of the Internet and related issues of surveillance, privacy and access.[38] Many of the issues at stake in this debate find resonance in the humanitarian sector. For while the impact of the Internet, mobile communications, Twitter, Facebook and a growing array of related sites has been positive in many ways, the opening up of these technologies has also created a new form of 'noise' that threatens at least some of the benefits.

Noise in an operational context. On 11 April 2012 an 8.6-magnitude quake occurred 610 kilometres southwest of Banda Aceh, Indonesia – an area that less than a decade before had experienced one of the most serious tsunamis in recent history. In responding to the perceived crisis, each of the countries in the region took a different approach. Some simply issued warnings while others ordered coastal evacuation. The Thai authorities shut down Phuket International Airport, for example, while the Chennai port in southern India was only closed for a few hours.

In the end, the quake did not generate a tsunami, but it triggered a considerable amount of chaos. In Sri Lanka, for example, coastal bus and train services were stopped, electricity was shut down and public offices were abruptly closed. Some journalists and activists tweeted for several hours, providing ground-level updates as well as relaying news from international wire services. In contrast, state agencies mandated to issue warnings relied on faxes and phone interviews with broadcast channels.[39]

This sort of information 'noise' can take many forms. According to a senior UN official in charge of coordinating the earthquake response in Haiti, one of the difficulties that he faced was the amount of discordant information that came through social media.[40] In one tragic case, a family sought assistance from a Search and Rescue Team to rescue their daughter who, as it transpired, had been dead for at least two days. Deploying the team was costly in terms of time and resources, but for the desperate family the rescue of their daughter was vital, even though they knew that she was dead.

The issue of 'noise' – that volume of messages and the cacophony of often contending information – can wreak havoc in efforts to assess needs and respond adequately. To some extent noise is inevitable. However, its consequences can be extremely costly in terms of human and financial resources, and, more importantly, in terms of lives and livelihoods. And yet there are means that can, to some extent, mitigate the effects of noise. There are, for example, means of weighing the plausibility of messages. Experience with mapping, for example, has enabled technicians to cluster open source information in ways that create patterns of impact and requirements. While by no means an all-encompassing solution, such mapping exercises add a degree of much-needed clarity.[41]

Taking this logic forward, it might eventually become a critical role for the UN – both in-country and at headquarters level – to serve as an information conduit and standard-bearer of humanitarian crisis-related information. This sort of role would require the UN system to hand over much of its relief-distribution role to the myriad of community, national and regional organisations with requisite operational capacities, and focus instead on the challenge of reconciling contending, inconsistent and complex information to save peoples' lives and livelihoods.[42]

The information admix. Often noise is generated because there is a lack of alternative information that leads people 'to fill in the blanks' from their own often very isolated perspectives. Here, as evidenced in the case of the Japanese tsunami in March 2011, the use of radio became a system that added focus and direction to people's Internet comments. In a related vein, and again drawn from the experience of the Japanese tsunami, another means of addressing the issue of noise is to recognise that all possible channels and technologies, from the highest to the lowest, must be used to ensure the sorts of focused protection that are needed in humanitarian crises.

Japanese citizens are major users of the Internet and social media networks, and Japan is media rich and very digitally enabled. And yet, when it came to understanding what was needed and what people should do to protect themselves from the combined impact of the tsunami and nuclear-related issues, unlike mobile telephony and the Internet, 'radio and print were less dependent on grid-band power supplies' and demographically one could not assume that everyone in a time of crisis was comfortable with or adept at social media.[43]

In other words, the full flush of social media's benefits to humanitarian action has to be tempered by a host of realities that are on the one hand specific to the present context and on the other more structural. In the case of the Japanese tsunami, it would seem that assumptions about the utility of social media had failed to take into account the fact that the elderly, particularly in rural areas, were frequently not engaged. This demographic reality was underscored by then UK Chief Scientific Advisor Sir John Beddington following the launch of the UK's 2012 Foresight study, 'Reducing Risks of Future Disasters'. Beddington noted that

> the generation born, who will have grown up with social networking and with a digital age, is starting to turn into adulthood and at the same time you have an older and more elderly population, which arguably could become partly disfranchised... As an ageing population and older workforce creates changing identities, inequality in digital knowledge and skills, particularly among older workers and the retired elderly, will need to be addressed.[44]

The demographic gap in the use of social media will inevitably adjust over time. Structurally, however, the utility of social media and its importance must also be considered in the context of disasters themselves. The growing complexity of disasters and their destructive force may well mean that the very structures and systems that keep social media functioning may also become 'victims' to disasters. The cybernetic systems and the fibre networks upon which so much depends are also vulnerable.[45] Yet this threat is offset by what is perhaps the greatest strength of the open source environment: the ubiquity of information, spread by means of a multitude of different and independent media, means that OSINT will always have something to contribute, whatever the nature of the disaster.

Notes

1. Personal communication between the author and a Syrian refugee now in London who worked for an international NGO and was arrested by the Syrian authorities in 2012.
2. For a comprehensive study of the history of humanitarianism, see Michael N. Barnett, *Empire of Humanity: A History of Humanitarianism* (Ithaca, NY: Cornell University Press, 2011).

3. John Cosgrove, *Synthesis Report: Expanded Summary Joint Evaluation of the International Response to the Indian Ocean Tsunami* (London: Tsunami Evaluation Coalition, 2007), p.9.
4. Ibid.
5. Ibid., p.12.
6. It is probable that the opinions of affected people have rarely, if ever, been so canvassed as they were in this disaster. In the past it has been less common for affected populations to be asked for their opinions about the aid they have received. This constant surveying of the views of the affected population may be one of the most significant innovations of the tsunami response. The majority of these surveys have not been carried out by implementing agencies seeking to know how well they have done, but by third parties, for academic study or for external evaluation.

    John Cosgrove, *Synthesis Report*, p.49.
7. See David Hulme and Michael Edwards, *NGOs, State and Donors: Too Close for Comfort* (London: Palgrave Macmillan, 2013). This is a revised version of their original work in 1997. It is interesting to note their own views of what had happened in this context between 1997 and 2013:

    > In 1997 we investigated the ways in which NGO-State-Donor relationships have changed role that NGOs play in development, asking whether their growing popularity had helped them to 'solve' the problems of poverty or had changed them to become part of the 'development industry' that they used to criticize. Using case studies of African, Asian and Latin American NGOs, we highlighted that the evidence suggested that NGOs were 'losing their roots' – getting close to donors and governments and more distant from 'the poor beneficiaries they sought to assist'. Since the book was first published, NGOs have continued to rise in number, scale and prominence, but our concerns have been little redressed and our argument remains strong today.

8. Private correspondence with me, London, 2 May 2013.
9. Eric Schmidt and Jared Cohen, *The New Digital Age: Reshaping the Future of People, Nations and Business* (London: John Murray, 2013), p.6.
10. Ibid.
11. This comment was made in discussions at a workshop of regional directors of a major European NGO in May 2012.
12. Richard Duncombe, 'Understanding Mobile Phone Impact on Livelihoods in Developing Countries: A New Research Framework', *Development Informatics Working Paper Series* (2012), Centre for Development Informatics, Institute for Development Policy and Management, University of Manchester, p.11.
13. 'Librarians, Internet Improve Farmers' Livelihoods in Romania', IREX, 15 May 2013, http://www.irex.org/news/librarians-internet-improve-farmers%E2%80%99-livelihoods-romania.
14. World Bank, 'Remittance Market Outlook', http://go.worldbank.org/K00UJL9JJ0.
15. Humanitarian Futures Programme, 'Private Sector Innovation and Humanitarian Action Workshop Report', 12 September 2012, p.7, http://www.humanitarianfutures.org.
16. BRAC was founded in 1972 as the Bangladesh Rehabilitation Assistance Committee. After a further change to its name, it was decided to just use BRAC as the full name of the organisation.

17. See World Bank (20 November 2012) press release announcing data at http://www.worldbank.org/migration.
18. It is interesting to note that the UK's Department of International Development in its Humanitarian Emergency Response Review called for greater collaboration with civil society, including the diaspora. 'Humanitarian Emergency Response Review', *Humanitarian Emergency Response Review Team*, 28 March 2011, https://www.gov.uk/government/uploads/system/uploads/attachment_data/file/67579/HERR.pdf.
19. The GSM Association (GSMA) 'represents the interests of mobile operators worldwide. Spanning more than 220 countries, the GSMA unites nearly 800 of the world's mobile operators with 250 companies in the broader mobile ecosystem, including handset and device makers, software companies, equipment providers and Internet companies'. See 'About us', GSMA, http://www.gsma.com/aboutus/.
20. Eric Schmidt and Jared Cohen, *The New Digital Age*, p.22.
21. Manuel Castells, *The Rise of the Network Society*, Revised Edition (West Sussex: John Wiley & Sons, 2010).
22. Saskia Sassen, *Global Networks, Linked Cities* (New York: Routledge, 2011).
23. This table was adopted from Jason Christopher Chan, 'The Role of Social Media in Crisis Preparedness, Response and Recovery', *Vanguard Report*, http://www.oecd.org/governance/risk/The%20role%20of%20Social%20media%20in%20crisis%20preparedness,%20response%20and%20recovery.pdf.
24. M-Pesa ('M' for mobile, *pesa* is Swahili for money) is a mobile-phone based money transfer and microfinancing service for Safaricom and Vodacom, the largest mobile network operators in Kenya and Tanzania. Currently the most developed mobile payment system in the world, M-Pesa allows users with a national ID card or passport to deposit, withdraw and transfer money easily with a mobile device.
25. Christian Aid, Christian Community Services Mount Kenya East, Kenya Meteorological Department, University of Sussex, UK Met Office, Humanitarian Futures Programme, King's College, London, 'Operationalising Climate Science: An Exchange between Climate Scientists and Humanitarian and Development Policy Makers: Kenya Demonstration Case Study', November 2012, http://www.humanitarianfutures.org.
26. Senegalese Red Cross, Senegal National Agency for Civil Aviation and Meteorology, Humanitarian Futures Programme, King's College, London, UK Met Office, University of Liverpool, Consultative Group on Agricultural Research, Climate Change, Agriculture and Food Security, 'Operationalising Climate Science: An Exchange between Climate Scientists and Humanitarian and Development Policy Makers: Senegal Demonstration Case Study', November 2012, http://www.humanitarianfutures.org.
27. Put simply, we've found that the people responsible for developing the services, programs, and policies that constitute social action – I'll refer to them as experts here – have a different frame of reference than the people they want to help. This limits the extent to which the experts can actually be of help. It limits what can be achieved through social action and even worse, sometimes leads to serious inadvertent harms.

Presentation by Dr Roz Lasker, Glen Cove Conference on Strategic Design and Public Policy coordinated by the United Nations Institute for Disarmament Research, the Center for Local Strategies Research, University of Washington, and the Said Business School, University of Oxford, June 9–11, 2010. See also R. D. Lasker, *Redefining Readiness: Terrorism Planning through the Eyes of the Public* (NY,

USA: New York Academy of Medicine, 2004), http://www.redefiningreadiness. net/pdf/RedefiningReadinessStudy.pdf; R. D. Lasker, *With the Public's Knowledge: A User's Guide to the Redefining Readiness Small Group Discussion Process* (NY, USA: New York Academy of Medicine, 2009), http://www.redefiningreadiness.net/pdf/ With_Public_Knowledge.pdf.
28. Jason Chen, 'The Role of Social Media in Crisis Preparedness, Response and Recovery', OECD, 2010, http://www.oecd.org/governance/risk/The%20role%20of% 20Social%20media%20in%20crisis%20preparedness,%20response%20and%20 recovery.pdf.
29. Remark made in meeting of NGOs attended by the author, London, March 2012.
30. For Haiti, remittances from the US account for the largest part of the national economy. US immigration officials were ramping up the deportation of Haitians who were living illegally in the US, when the devastating earthquake struck Port-au-Prince in 2010. In response to the disaster, the Obama administration granted Haitians living in the US illegally or those with expiring visas 'temporary protected status'. The designation allowed about 200,000 Haitian immigrant workers to remain in the country and to continue sending remittances back home.
31. Warren Miller et al., *Reaching Out to Those in Need: A Guide to Establishing a Successful Disaster Relief Fund* (NC, USA: Z. Smith Reynolds Foundation and the Duke Energy Foundation, 2005), p.17.
32. Stephen Morris and John Horrigan, '13 Million Americans Made Donations Online After Hurricanes Katrina and Rita', *Pew Internet*, 24 November 2005, http:// www.pewinternet.org/Reports/2005/13-million-Americans-made-donations-onli ne-after-Hurricanes-Katrina-and-Rita.aspx.

> At the same time, a notable number of Americans used the Internet to move beyond traditional institutions as they looked for news about the hurricanes and information about relief efforts. Some 17% of those who got news about the disasters read blogs to get details and insights into the impact of the hurricanes on affected communities. Some 5% of Internet users – or 7 million people – went online to set up their own relief efforts. And 4% posted their own material such as comments, links, and pictures related to the hurricanes on online blogs, bulletin boards or chat rooms.

33. Sue MacGregor, 'The Reunion: Hurricane Katrina', BBC Radio 4, 29 August 2010, http://www.bbc.co.uk/programmes/b007x9vc/episodes/2010.
34. Stacey White, *Turning Ideas into Action: Innovation within the Humanitarian Sector: A Think Piece for the HFP Stakeholders Forum* (UK: Humanitarian Futures Programme, November 2008), p.4.
35. See, for example, Danish Refugee Council and the Community-Driven Recovery and Development Project, 'Piloting Accountability Systems for Humanitarian Aid in Somalia', *Humanitarian Innovations Fund*, ELRHA, June 2011, http://www. humanitarianinnovation.org/projects/large-grants/drc-somalia.
36. 'Aid Policy: From Rwanda to Haiti – What Progress on Accountability?', *IRIN*, 4 July 2012, http://www.irinnews.org/report/95780/aid-policy-from-rwanda-to-haiti-what-progress-on-accountability.
37. Ibid.
38. Mark Mazzetti and Michael S. Schmidt, 'Ex-Worker at CIA Says He Leaked Data on Surveillance', *New York Times*, 10 June 2013, http://www.nytimes.com/2013/ 06/10/us/former-cia-worker-says-he-leaked-surveillance-data.html.

39. The substance of this case comes from Rohan Samarajiva and Nalaka Gundwardene, 'Crying Wolf Over Disasters Undermines Future Warnings', *SciDevNet*, 6 February 2013, http://www.scidev.net/global/policy/opinion/crying-wolf-over-disasters-undermines-future-warnings-.html.
40. Private correspondence with me, March 2011.
41. In 2009, crisis-mapping consisted of at least 1,500 organisations in 120 countries. In 2009, according to Crisis Mappers: The Humanitarian Technology Network, http://crisismappers.net, there are over 3,000 organisations working in 162 countries. In the aftermath of the March 2011 Japanese tsunami, OpenStreet Map had a volunteer community that mapped over 500,000 roads that had potential operational access, and Sinsai.info categorised and mapped 12,000 tweets and e-mails from the affected. For more detailed analysis, see Patrick Meir and Jennifer Leaning, *Applied Technology and Crisis Mapping and Early Warning in Humanitarian Settings* (Cambridge, MA: Harvard Humanitarian Initiative, 2009), http://hhi.harvard.edu/sites/default/files/publications/publication%20-%20crisis%20mapping%20-%20applying%20tech.pdf; and John Crowley and Jennifer Chan, *Disaster Relief 2.0: The Future of Information Sharing in Humanitarian Emergencies* (UN Foundation & Vodafone Foundation: UN OCHA, 2011), http://issuu.com/unfoundation/docs/disaster_relief20_report.
42. Changing the UN's fundamental humanitarian roles and responsibilities has been an issue of considerable debate over the past decades. In Dalton et al., *Changes in Humanitarian Financing: Implications for the United Nations* (New York: UN OCHA, 2003), the authors argue that the UN system will have to change its approach to humanitarian assistance fundamentally. In the first instance it will have to limit its distribution activities which in so many ways places it in competition for resources with other humanitarian organisations, and instead use its authority and international role to add veracity to the sorts of data which, as a coordinator – not in operational competition with others – it should be able to do.
43. Lois Appleby, *Connecting the Last Mile: The Role of Communications in the Great East Japan Earthquake* (London: Internews-Europe, 2013), p.9. The author goes on to mention that in light of the power-supply problems that affected many aspects of the high-technology communications systems, newsletters – even hand-written newsletters – at the local level became fundamental to exchanging information about the crises.
44. Sir John Beddington on BBC Radio 4, *Today Programme*, 21 January 2013, regarding the launch of the Government Office for Science, *Reducing Risks of Future Disasters: Priorities for Decision-Makers* (London: Government Office for Science, 2012), http://www.bis.gov.uk/assets/foresight/docs/reducing-risk-management/12-1289-reducing-risks-of-future-disasters-report.pdf.
45. It is worth noting that when Google introduced Google Balloon in June 2013 in an effort to reach remote areas via near-space transmissions, one justification for the initiative was to deal with areas that had lost communication capacity due to disasters. Leon Kelion, 'Google Tests Balloons to Beam Internet from Near Space', *BBC News*, 15 June 2013, http://www.bbc.co.uk/news/technology-22905199.

7
Human Security Intelligence: Towards a Comprehensive Understanding of Complex Emergencies

Fred Bruls and A. Walter Dorn

Introduction

Humanitarian crises may arise from natural disasters, such as droughts, floods and earthquakes, or they may be caused or exacerbated by human beings through armed conflict. The latter are often referred to as 'complex emergencies'. These emergencies call for holistic international responses that need to be coordinated across the variety of humanitarian and military actors. The responses require a great deal of information, situational awareness and occasionally secret intelligence. But the information contained in open sources usually provides ample basis for those organisations seeking to respond positively to these crises.

This chapter seeks to provide an overview of a new and much needed intelligence concept, Human Security Intelligence (HSI), the reasons for its utility and some of the issues arising from its use.[1] It will begin by providing an overview of the international operations that seek to deal with complex emergencies, namely peace operations (POs), before analysing some of the current, flawed and incomplete intelligence practices. The new intelligence concept, HSI, will be introduced as a means to develop a broader understanding of the environment, an understanding that is centred on the civilian populace. OSINT, it will be argued, is a key information source for the HSI model. The production of HSI requires a multidisciplinary and multiagency approach. In particular, military forces need to work with other agencies to develop effective responses based on strong civil-military cooperation. From this new paradigm arise many questions: How should HSI be managed to enable optimum collaboration? How should information be prioritised in this complex and information-rich environment? The chapter will seek to elaborate on these and other questions, and suggest how HSI can make POs more effective, especially as they have evolved to undertake enormous mandates in the twenty-first century, when the challenges remain daunting.

Complex intelligence and evolving peace operations

Complex emergencies have been defined by the International Federation of Red Cross and Red Crescent Societies (IFRC) as a state of 'total or considerable breakdown of authority resulting from internal or external conflict and which requires an international response that goes beyond the mandate or capacity of any single agency and/or the ongoing UN country program'.[2] The IFRC goes on to characterise a complex emergency as involving[3]:

- extensive violence and loss of life;
- displacements of populations;
- widespread damage to societies and economies.

The Red Cross also highlights key issues of concern in the humanitarian response:

- the need for large-scale, multifaceted assistance;
- the hindrance or prevention of assistance by political and military constraints;
- significant security risks for relief workers in some areas.

Complex emergencies involve humanitarian actors, such as relief agencies on the ground, and often necessitate the creation of peace operations, especially following the conclusion of a peace agreement or tentative understanding between the conflicting parties.[4] These POs are launched by the UN or regional organisations in order to create a safe and secure environment for the local populace and to help to build the physical and social infrastructure required for a sustainable peace. With such a complex mandate, these operations must coordinate their actions with and among a range of actors, especially the conflicting parties and those involved in local governance. POs are an important means of working towards an integral solution to complex emergencies.

In today's POs, military forces are confronted with new concerns in comparison with more traditional peacekeeping operations, which were usually located on state borders to separate standing armies. Current conflicts, by contrast, are characterised by:

- an increasing number of non-state actors[5];
- ethnic, religious or cultural disputes throughout large regions or countries;
- asymmetry as opposed to symmetry in the combatant forces – for example, insurgents versus a government; insurgents are often not distinguishable from the local populace in which they seek sanctuary; their modus operandi can include terrorism and the use of proxy forces[6];

- the need for international interveners to secure the 'goodwill' and support of the populace; this is the new 'centre of gravity' for modern POs[7]; this goodwill is based on the perceived legitimacy of the intervention force; actions deemed inappropriate for the local populace, especially 'collateral damage' or civilian casualties, can rapidly erode this perceived legitimacy;
- diversity of intervening actors, not only military peacekeepers in the field, but including a large number of international (UN) agencies and NGOs; since unity of effort between all of these actors is a key for success in humanitarian operations, coordination and negotiation are extremely important daily activities[8];
- privatisation of the security sector, in overseas operations and in the host country; private security companies (PSCs) have become significant players in the international operations and even for the combatants; estimates put the total value of the private military and security industry at USD 210 billion in 2010[9]; their clients include both states, international organisations and NGOs, that contract out specific tasks, such as the protection of compounds, buildings, convoys and personnel[10]; local authorities also often hire PSCs for their own protection; PSCs also employ large numbers of local personnel, giving them both influence and intelligence at the local level; for example, it is estimated that in Afghanistan some 90 PSCs are active, employing over 20,000 Afghan personnel.[11]

The increased number of actors and complexity of armed conflicts has led to a significant evolution in POs. More traditional peacekeeping operations, such as the UN Peacekeeping Force in Cyprus (UNFICYP) or the Multilateral Force and Observers (MFO) in the Sinai, took place in a predefined area of responsibility (AOR), consisting of a buffer or demarcation zone between the belligerents. These parties had usually retreated behind fortified lines and had agreed to the presence of the peacekeeping forces. The AOR was mostly uninhabited, due to evacuations of the civil populace. The security of the local populace within the AOR was therefore of less importance.

The more complex peacekeeping operations of the twenty-first century have to deal with the multifaceted problems of complex emergencies. Experience has taught the world that, although the belligerent parties may have agreed to a ceasefire and a peacekeeping force, they also often breach and renege on their agreements. Much more is needed to secure the peace: a functioning society that offers strong political, economic and social alternatives to war. Belligerents also need to be effectively deterred from returning to active hostilities.

Because of this, modern peace operations have largely abandoned the restrictive Rules of Engagement (ROE) found in traditional peacekeeping operations. They have become more robust, including larger forces, such as the Intervention Brigade in the United Nations Stabilization Mission in the Democratic Republic of the Congo (MONUSCO). In these more complex peacekeeping operations, peacekeeping forces are deployed countrywide instead of on state borders or in no-man's land.[12] Reflecting these changes, new terms have been introduced to describe the various evolving tasks of POs, including peacemaking (negotiations for peace); peace-building (developing the physical and social infrastructure for a sustainable peace); peacekeeping (providing security); peace enforcement (to apply force to stop or reverse of aggression); and preventive deployment (to stop aggression before it occurs).

The consent of all belligerent parties, although an important prerequisite for the initial deployment of a UN peace operation, no longer blocks intervention when circumstances demand the protection of civilians (POC) or enforcement of UN Security Council resolutions.[13] Increasingly, POs use armed force to deal with troublesome 'spoilers' of the peace process. Examples of robust peacekeeping and peace-enforcement operations with robust mandates can be found in Mali, Côte d'Ivoire and the Democratic Republic of the Congo. Although the NATO operations in Afghanistan cannot be labelled as POs, such counterinsurgency operations share common characteristics with POs, and hence can be useful for extracting lessons. Conversely, the UN Assistance Mission in Afghanistan (UNAMA) is an example of a small PO that may in the future expand but has not yet deployed an armed force of its own.

Given the complexity of the environment in which POs take place, and the multiplicity of actors involved, many different facets must be taken into account when planning for missions within the AOR. Current and accurate intelligence is required to understand this environment and for the planning, execution and evaluation of POs. The intelligence branches of the POs are responsible for providing an understanding of the background and current situation in the AOR, and for providing the mission leaders with a common operational picture (COP). According to military doctrine, the COP constitutes a 'snapshot in time' of the mission environment and all military forces present (friendly, hostile and neutral) therein. The COP contributes situational awareness to the commanders, which is their understanding of the mission environment in the context of their mission.[14]

Intelligence models used in the past and present are, however, insufficient to provide a comprehensive understanding of complex emergencies. Given that warfare has moved from predominantly rural to urbanised terrain, the intelligence effort must shift from the 'physical terrain' to the 'human terrain'. The populace has become an extremely important factor for operational success of a peace operation.[15] Which parties do armed persons belong to, which groups do they support and why? What are the threats that confront the populace? What indigenous knowledge, traditions and systems can contribute to a sustainable end state that is acceptable to the various parties and groups? The populace in wartorn countries is traumatised, an aspect mutually shared by all the communities and groups.[16] A new paradigm offers the possibility of increasing the scope and effectiveness of intelligence and hence the perceived legitimacy of the overall operation. Unfortunately, most approaches in these operations, whether they are taken by the military, the UN or NGOs, are self-centred, resulting in a fragmented picture and stovepiped mission approach, thereby eroding the mission's legitimacy in the eyes of the populace.[17] For instance, resorting solely to military force is considered to be one of the worst options to influence behaviour. More intelligent and intelligence-driven means are required.

The concept of human security intelligence

Because of the multidimensional nature of modern conflicts and of international responses, those taking part in POs need much stronger information and intelligence-gathering capabilities across the full spectrum of human activities. Because peace is holistic, the mission must also be. A comprehensive understanding of the total operational environment is needed across all domains, not solely the military domain. Still, intelligence professionals must prioritise the most relevant information. To do this, various intelligence models are currently being operationalised by Western forces but these are inadequate.

In traditional war-fighting and peacekeeping operations, military actors focused almost exclusively on threats to physical security, which usually meant watching or targeting opposing military forces. But in the present day's enlarged agenda, the threats are not merely posed by armed attacks against one's own forces; they are also psychological, political, economic and cultural threats to local civilians. To the goal of 'freedom from fear' must be added 'freedom from want', allowing the population not only to survive but also to thrive so that peace can be sustained. The interconnection between these two goals necessitates an interconnected framework, as shown in Figure 7.1.

Given the relationship between the freedom-from-fear and freedom-from-want dimensions, the Common Operational Picture (COP) must cover all factors of human security, putting the civilian populace at the centre.[18] A large range of factors influence events in the AOR and its direct surroundings. Because of this a larger area of interest (AOI) needs to be defined to encompass them.[19] One of the main tasks of intelligence staff – military, police and civilian, whether combined into one organisation or not – is to provide the mission leader with a detailed yet prioritised COP at many levels.

The intelligence models which are most frequently applied in current Western military operations are insufficient to analyse the human security situation in an AOI. The HSI model, an alternative to existing models, is presented, based on the broad definition of human security, as provided by

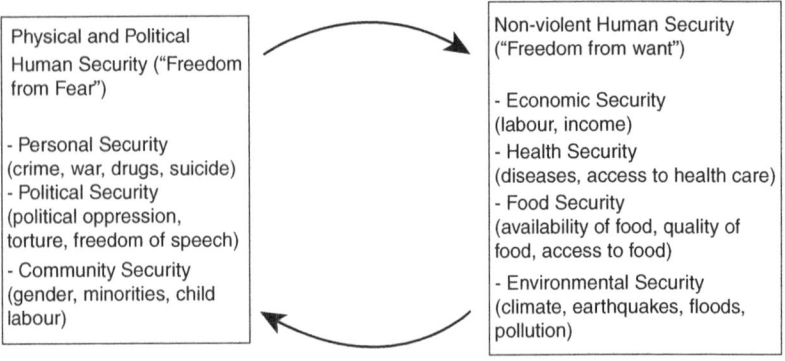

Figure 7.1 Causal pathways of human security[20]

the 1994 *Human Development Report*.[21] If correctly tested and applied, this model could serve as a key early warning mechanism to help to alert humanitarian actors and allow them to prevent the escalation of complex emergencies as well as to resolve them. The existing models needs much improvement.[22]

Intelligence models in today's operations

Most intelligence staff of Western militaries use various combinations of the models set out in US Army doctrine in order to develop an understanding of the operational environment. These US models are mostly based on so-called 'instruments of national power' that can be used to achieve 'theatre, national and/or multinational objectives'.[23] These intelligence models, which list the principle factors to be analysed, are known by acronyms like DIME, DIMEFIL, ASCOPE and PMESII.[24] Today, DIME and DIMEFIL, though still well known in military circles, are hardly used in POs.[25] The others are still in frequent use and are worth summarising.

The ASCOPE model is currently used in some peace and stabilisation operations. For example, this analytical tool was used by the UK and Australian intelligence communities in Afghanistan. The ASCOPE model analyses the civil aspects[26] of the AOI in the following dimensions[27]:

- areas: analysis of the influence of key civilian areas on military operations and vice versa;
- structures: analysis of physical infrastructure, such as buildings, bridges, roads, railways and communication towers; also the presence of possible toxic materials is taken into account;
- capabilities: analysis of the capabilities required and present in the AOI to save, sustain and enhance life, such as public administration, food, emergency services and health care;

- organisations: analysis of presence, activities and organisational composition of non-military groups and institutions in the AOI with respect to their influence on the populace, the military mission and vice versa;
- people: analysis of non-military people in the AOI in terms of opinions, actions and political influence;
- events: analysis of events in the AOI that affect the populace, military operations, non-military organisations, religious and national holidays, crop harvest and elections; also unplanned events such as civil unrest, environmental or natural disasters, and industrial accidents are taken into account.

The US Army Field Manual 3-0, which sets out the fundamentals of warfighting, focuses strongly on the PMESII model for the variables affecting operations.[28] Because of this, PMESII has become a standard in peace and stabilisation operations conducted by NATO countries, although the model was not originally intended for adoption in these operations.[29] For example, in Afghanistan, PMESII was used as the intelligence standard for the US and Dutch forces.[30] The PMESII model can be used to analyse the operational side of the AOI.[31] The dimensions explored by this model are[32]:

- political: analysis of political organisations, groups and individuals and their linkages within the AOI;
- military: analysis of military organisations and their capabilities in the AOI;
- economic: analysis of the economic position and health of groups in the AOI;
- social: analysis of social networks within and between groups and social links of individuals;
- information: analysis of the information position of a group in terms of the information known by the group, use of information within the group, propaganda, news media and so on;
- infrastructure: analysis of the infrastructure within the AOI in terms of roads, railways, airports, power supply, sanitation, capabilities like governance organisations and so on.

Often the basic PMESII model is extended with a number of extra dimensions, such as physical environment, time, crime, narcotics or others relevant to the specific mission. Philosophically, this model views the operational environment as a system of interacting subsystems. For example, ethnic or religious groups are viewed as separate interacting systems. These groups are further composed of tribes that are considered to be systems in their own right. Terrorist groups operating in the AOI are also considered to be influencing systems. This philosophy, referred to as the 'system of systems approach', provides the intelligence community with a powerful instrument with which to analyse the AOI.[33]

Shortcomings of the present intelligence models

The effectiveness of the aforementioned intelligence models in humanitarian and peace operations is, however, limited by a number of factors. First, the models were developed to influence opposing parties or regimes, rather than to address the root causes of crises. The models are enemy-centric or military-centric rather than population-centric. While these factors may be important in the planning process, they are not necessarily linked directly to the desired end state (the security, peace and wellbeing of the population) and thus give only partial guidance to achieve that end state.

A related problem is with the application of the models. They are linear models, meaning that analysis is conducted along a fixed line. For example, in the case of PMESII the line starts with the 'political' dimension and ends with the 'infrastructure' dimension. Admittedly, intelligence analysis often consists of sorting gathered information into the various dimensions. Reality, however, is not linear and not so easily categorised. Linear models limit the flexibility of the analysis and can create an impression of 'static variables within static frameworks' while in reality the operation needs to be dynamic to succeed.[34] This can cause the model and its variables to be misapplied in different circumstances where the context is very different, notably peacekeeping instead of war-fighting.

US Army Field Manual 3-0[35] extends the PMESII model by adding physical environment and time (PMESII-PT) to include more context in the model for intelligence analysis. However, even extending PMESII in this way does not lead to a model providing a more dynamic understanding of the situation, since the model still requires that gathered information is sorted into different dimensions. Area and time are considered as discrete dimensions, and the connection of these with the others (within PMESII) is not analysed.

The second problem in the application of the standard PMESII-type models is that it does not require analysts to consider the broader context. Dimensions like culture, demographics and others go unanalysed, though some of these may be encompassed in the social dimension.[36] A consideration of the broader context is needed to create a holistic picture of the operational environment.

A truly holistic analysis will need to encompass narratives (storylines) in addition to the tabular boxes (cells in a database) typical of linear models. An approach that is holistic in nature is not implemented through simple fact-finding alone, but also by placing these facts in their context, such as time, location and culture or whatever gives the most meaning to the facts. In this respect, some practitioners have distinguished between information (dealing with the 'what'), knowledge (dealing with the 'how') and understanding (dealing with the 'why').[37]

Whereas a linear model is usually based on, and too often ends with, quantifiable data that are fed into the various dimensions, a holistic approach results in a freer interpretation of the facts.[38] Naturally, interpretations are influenced by the values and beliefs of the analyst.[39] So a key factor for a realistic analysis is the ability of the analyst to grasp the background cultural and historical narrative. As a well-known military anthropologist has noted, it is clear that 'misunderstanding culture at a strategic level can produce policies that exacerbate an insurgency; a lack of cultural knowledge at an operational level can lead to negative public opinion and ignorance of the culture at a tactical level endangers both civilians and troops'.[40]

Human security intelligence as a model

The aforementioned intelligence models, widely used by NATO intelligence analysts, do not effectively address all dimensions determining 'human security'.[41] The UN Development Programme's (UNDP) highly influential *Human Development Report 1994* introduced the concept to encompass the broader dimensions of security. The factors presented in this original definition of human security are characteristic of a people-centric approach. As shown in Figure 7.2, there is a direct link between the various dimensions of human security. Ignoring one or more of these dimensions can lead to unforeseen risks for the populace as well as the intervening forces and other humanitarian actors.

Therefore an intelligence model needs to be developed to analyse the human security situation within the AOI. The UNDP components of human security are personal security, community security, political security, economic security, food security, health security and environmental security.[42] The model and some examples of the various components are described below.

Personal security is freedom from physical violence. According to the *1994 Human Development Report*, in both poor and rich nations, human life is increasingly threatened by sudden and unpredictable violence. The report mentions several forms of threat, including[43]:

- those from the state (physical torture);
- those from other states (war);
- those from other groups of people (ethnic tensions);
- violence between individuals or gangs (crime, street violence);
- violence against women (rape, domestic violence);
- those against children based on their vulnerability and dependence (child abuse);
- violence against the self (suicide, drug use).

Community security is freedom from oppressive actions within a community. Specific examples cited by the report include acts of oppression, such as slavery, harsh treatment of women and discrimination, particularly within

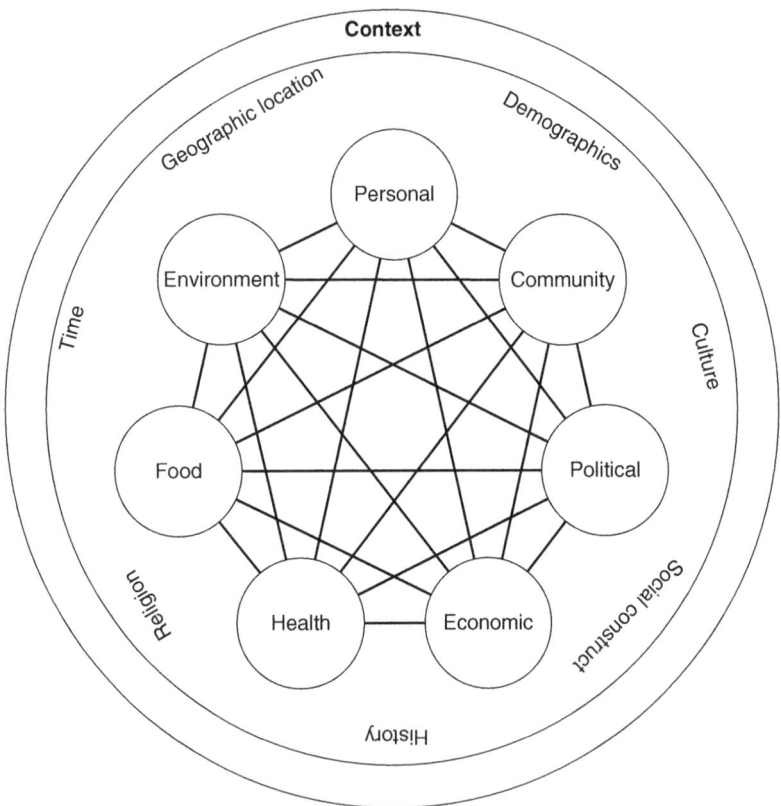

Figure 7.2 Human security intelligence model

more traditional communities. The report also cites the disappearance of traditional languages and cultures as a threat.[44] These practices can culminate in direct threats to personal security – for instance, through tribal wars or ethnic cleansing.

Political security is freedom from state repression. The *1994 Human Development Report* gives the following examples of state repression: political repression, political detention and imprisonment, systematic torture, disappearances, and control of ideas and information.[45]

Economic security means, among many things, an assured basic income. Productive and remunerative work, or in the absence thereof a public financial safety provision, is key to economic security.

Health security can be measured by human life expectancy, absence of diseases and access to health care.

Food security relies on physical and economic access to basic food and water.

Environmental security relates to the general condition of the environment and effects of environmental disasters on human lives.

Following the causal pathways depicted in Figure 7.2, personal, community and political security can be regarded as subsets of freedom from fear, while economic, food, health and environmental security are subsets of freedom from want.

Holistic models, which result in a richer understanding of the environment, are more difficult to summarize using acronyms since they are constructed on case-specific combinations of relevant dimensions. Given the fact that human security is a holistic construct, the temptation to create an acronym for the elements of the HSI model is best left unsatisfied.[46] Rather, the various security types outlined by the UNDP should be analysed in their relevant context to give true meaning to their variables. The model emphasises that not only do the variables within the different dimensions matter, but more importantly the causal relationships between the dimensions matter. Furthermore, the aforementioned problem that analysts often do not sufficiently take context into account is addressed by making the context part of the model itself.

Admittedly, the proposed HSI model is a complicated one, creating a multidimensional picture that can be difficult to conceptualise simply. The number of variables influencing the various human security types considered in the model is almost endless. The savvy analyst will have to prioritise the factors that are most pertinent. Given the large number of causal pathways, a true holistic analysis can be conducted in the form of a narrative to complement or replace a database of variables.[47]

The value of the proposed HSI model lies in the fact that it is population-centric or human-centric, and thus more directly related to the desired end state. When properly applied, it could provide valuable intelligence to guide an operation. It can also provide early warnings about possible conflicts to the commander of the PO. For example, in Haiti, food riots were triggered by global commodity prices, something that was not foreseen by the UN mission, whose intelligence efforts were focused on in-country indicators. Using the comprehensive approach afforded by HSI, the commander could have stood a better chance of identifying and addressing the root causes proactively and preventively rather than dealing with the resulting unrest.

Since the HSI model is constructed around the UNDP model, it does, however, not explicitly provide information regarding military and legal aspects which are emphasised in more traditional models. Therefore the model could be used in combination with other operational intelligence models, such as PMESII and ASCOPE.

Incorporating intelligence sources and OSINT

Having seen what a HSI model might be composed of, and some of its benefits, the next step is to consider how the various sources of information or intelligence can be incorporated into it. In the traditional intelligence literature, a number of collection disciplines or source types are distinguished:

- ACINT (acoustic intelligence): detection and tracking of sound;
- HUMINT (human intelligence): information from human sources (for example, talking or messaging with people);
- IMINT (imagery intelligence): images (still or video), such as photographic, infrared, multispectral, taken from platforms on the ground, in the sky or in space;
- MASINT (measurement and signature intelligence): scientific and technical information to identify equipment (for example, weapons) used in the AOI;
- RADINT (radar intelligence): information gathered by radar systems, for instance, to detect movement;
- SIGINT (signals intelligence), which can be divided into COMINT (communications intelligence) for the detection and tracking of communications from an individual or groups and ELINT (electronic intelligence) to detect and analyse electronic signals other than communications signals in the AOI;
- OSINT (open source intelligence): information available in the public domain, such as Internet, books and newspapers.

The intelligence sources used may be an outcome of both the environment and institutional preferences. For example, in a counterinsurgency operation where the local population may be hostile, as was often the case during NATO's operations in Afghanistan, the military alliance has leaned heavily on traditional means of information gathering, such as SIGINT, IMINT and MASINT. HUMINT is also valuable but sources need to be carefully vetted. Because modern POs tend to occur in HUMINT-rich environments, it is more natural for them to utilize this source. In Haiti, the UN mission went as far as to hire paid informants to gather information about gangs and their leaders to facilitate arrest operations.[48]

More generally, POs can rely heavily on OSINT since the activities of the mission are not secret. In wars and counterinsurgencies, adversaries go to great lengths to hide information from each other. In POs, the environment is more permissive, though not entirely. While OSINT is often the most important source for HSI, information found in open sources will need to be corroborated with information gathered directly in the crisis area, some of which may need to be gathered from secret sources and by secret means.

As explored in the other chapters of this volume, recent developments in IT have had huge impacts on the utility of OSINT. For instance, in Chapter 6, a range of these technologies have found uses in humanitarian contexts. However, the usual caveats regarding misinformation apply. So the savvy intelligence analyst learns how to find the best sources and how to corroborate information across as many sources as possible.

Gathering information in an exploding OSINT domain

With the information explosion covering many areas of human security, a major problem arises: one can easily drown in the available information. In humanitarian and peace operations, accurate and timely information is paramount for planning, execution and evaluation. The intelligence analyst, sometimes located in the field and sometimes elsewhere entirely, does not have the time to search through endless sources of information. Basic Internet search tools are valuable, but they also lead to large numbers of results that exceed the ability and the willingness of even a dedicated user to look through.[49] More specific search platforms such as Google (Scholar, Books, site-specific searches) and the Internet Archive (Wayback machine) can be powerful aids.

Prior to the establishment of a specific PO, useful information can be found directly on websites of international organisations, NGOs, governments and other organisations already active in the AOI. One problem here is that information can be poorly presented or manipulated. For example, indicators and indices can be, and sometimes actually are, misused by, for instance, aid donors, international investors, and even analysts and academics.[50] Aid donors sometimes exaggerate death tolls or the severity of a given situation for fundraising purposes.[51] The *Human Security Report 2009/2010*, for instance, mentions the excessive death toll estimate of the International Rescue Committee (IRC) of the war in the Democratic Republic of the Congo in the period 1998–2007. The report claims that this estimate was based on questionable methodology.[52] This underscores the conclusion that information from sources should be corroborated by others.

Given the fact that many sources manipulate information for a variety of purposes, just collecting and analysing the information found in open sources is not sufficient. Information about the sources found and used should also be recorded. For example, the ownership of the source, information about possible partisanship, track records regarding accuracy and so on should always be taken into account when information is being used.

For the purpose of monitoring social media such as Twitter, Flickr and Facebook, a growing number of tools are available for the intelligence community. Some of these methods and their implications are discussed elsewhere in this volume. However, a prerequisite for effective and timely OSINT collection, across the board, is the availability of a robust ICT infrastructure, with a good connection to the Internet and major databases (governmental and non-governmental). This should be borne in mind when intelligence cells are established since connectivity can pose a problem in conflict zones.[53]

From information to intelligence in the field

In militaries around the world, the processing of information at the brigade level typically takes place in a collation unit where collected information is stored at an all source intelligence cell (ASIC). There military analysts work to analyse and integrate this information in order to produce an assessment of the AOI, thus turning information into intelligence.

The UN has taken a more holistic approach in its operations since 2006: UN POs have a Joint Mission Analysis Centre (JMAC), usually based at the mission headquarters and sometimes at regional offices of the mission, where military, police and civilians work together. However, given that military personnel are much more numerous in peace operations and that military intelligence officers are easier to acquire, the JMACs tend to be mostly military. The JMAC looks at longer-term analysis while a Joint Operations Centre (JOC) looks at the daily intelligence needs of the mission.

The introduction of HSI in the operational environment is likely to cause some problems, especially since analysts will be confronted with much more information relating to many more dimensions than previously. Much of this information will come from the civilian domain, including open sources, but given the multiple causal linkages with the military domain, it must be handled as equally relevant as traditional intelligence. HSI analysis is multidisciplinary, and necessitates the engagement of expertise not routinely present in armed forces' intelligence communities. Therefore ASICs should be augmented with experts in the various domains, who in most cases must be drawn from civilian bodies, such as governmental or academic organisations.

Beside the challenges presented by this novel approach, fresh opportunities also arise, especially relating to new tools. Increasing computer power and new software solutions enable the JMACs, JOCs, ASICs and the intelligence community operating remotely to manipulate available information, visualise dynamics, and recognise patterns and indicators of trends.

Another key problem lies in information management, a key function that is often neglected or overlooked in military operations, since 'neither analysts nor collection managers/specialists want to be "information managers" and therefore responsible for the tedious naming, storing, archiving, organising, cross-referencing and retrieval of information'.[54] Too often information storage is regarded as an IT function, and thus left to the IT staff to deal with.

For example, during the first six rotations of the Task Force Uruzgan in Afghanistan, there was no information-management position within the intelligence staff.⁵⁵ No data-management protocols had been established, with all incoming and produced information stored in an unstructured folder system that was only partially searchable. On top of this, every rotation of the intelligence staff reinvented the entire information folder structure. This led to a situation where the intelligence staff of the seventh rotation of the task force was confronted with a system where the information was buried in an unmanaged and unstructured storage system consisting of over 42,000 folders that went up to 18 layers deep. Despite the presence of a word-search function, a large part of the information remained irretrievable.⁵⁶

The Australian contingent in Task Force Uruzgan used a different organisational approach to collation and analysis. The intelligence cell was composed of a 'front office' that consisted of only two or three officers, whose task was to collect the incoming information and relay it to an intelligence 'back office' in Australia. This Australia-based office was a permanent collation ASIC with staff serving longer than six months, thus gaining the necessary skills and experience for thorough analysis. The back office conducted analysis and collation in Australian databases, and sent intelligence reports to the Australian front office in Afghanistan for further dissemination. This ensured continuity within the workforce where lessons learned were directly fed into the organisation, thus leading to a constantly improving quality of work and intelligence output.⁵⁷ Such a model could be equally beneficial for other small armed forces, such as the Dutch and Canadian militaries.

HSI: Implications, challenges and opportunities

The increasing need for HSI will further increase the importance of OSINT, which is already one of the most important sources of information in complex emergencies. Much of the information for the HSI model can be gleaned from open sources, especially from governmental and non-governmental organisations.

Some NGOs in the field, however, may be reluctant or unwilling to contribute to intelligence cells since they feel that this could seriously jeopardise their neutrality.⁵⁸ So intelligence officials must be sensitive when they reach out to cooperate with NGOs.⁵⁹ In fact, some intelligence personnel act as 'salesmen', convincing specialists and organisations to cooperate closely with them.⁶⁰ Civil-military cooperation (CIMIC) officers can also play an important role to persuade civil organisations and individuals to cooperate. For example, in the Netherlands, CIMIC reserve officers were able to collect large quantities of valuable information for the new police-mentoring mission in the Afghan Kunduz province, mainly from open sources and from Dutch offices of NGOs already present in Kunduz. Ideally, all of the partners in peace operations would train together and adopt an information-sharing model, perhaps across the eight 'tribes' described by Steele (academia, civil society, commerce, government, law enforcement, media, military and non-government/non-profit personnel).[61]

The constraints provided by organisational cultures may also need to be overcome. For example, on the military side a shift of culture from 'need to know' to a 'dare to share' basis would be welcome. As already stated, civilian organisations in the mission area should be seen as potential new 'clients' for the products of the intelligence staff. As stated earlier, the intelligence community is quite reluctant to share information with these organisations, which are reluctant to share with the intelligence community. NGOs are, generally speaking, more willing to cooperate with the intelligence cell when they can expect something in return – for instance, security assessments or threat warnings, assessments on the fairness of election polls, or other kinds of information.[62] This also means that the introduction of HSI may necessitate a reassessment of classification levels.

Another possibility to fill the expertise gap is currently being examined by the Royal Netherlands Army, where an environment cell is taking the intelligence role. This cell consists of military staff that can be augmented with civilian staff, who have specific expertise that is normally not available within a military intelligence cell, such as economists, development specialists, medical specialists and environment specialists.[63] Local and regional expertise might also be added to this cell. However, civil experts are often reluctant to be sent to high-risk mission areas. Where, at mission level, the analysis component of the intelligence cell is often located at coalition headquarters in safer locations, such as at Supreme Headquarters Allied Powers Europe in Belgium, this reluctance may not play an important role.

Similarly, in Afghanistan, the US also worked to expand the people collecting information, to include anthropologists, as part of its Human Terrain System project. These academics contributed to a better understanding of the deeper cultural, linguistic and societal factors, and shared their insights with US and international forces. However, this caused considerable controversy: the American Anthropological Association, for instance, disapproved of using academics as intelligence-gathers and analysts since the information could be (and likely was) used to help to carry out combat operations against certain indigenous groups.

The use of civilian experts in the field could pose other problems.[64] The Law of Armed Conflict does not recognise these experts as combatants and as such they are not legally protected as a soldier is under the Geneva Conventions. Insurance can also play an important role, with many health insurance policies excluding armed conflict. Some militaries deal with this by including persons with the requisite skills in their reserve forces.[65]

A further organisational solution could be to outsource specific intelligence tasks to military support firms. These are a type of private security company with a number already being active in the field of intelligence.[66] However, these firms work on a commercial basis and working with them requires contracts and constraints. In the situation of Task Force Uruzgan, a UK military support firm was hired by the Dutch government to operate leased unmanned aerial vehicles (UAVs). The contract provided for a fixed number of flight hours, and the lack of flexibility regarding additional hours potentially led to a loss of valuable information. Also the effectiveness of flight hours was not discussed, which, due to technical problems with the UAVs, led to major expenditures for limited overall effectiveness.[67] Besides this, most military support firms working in the intelligence field are not specialised in the freedom-from-want dimensions of HSI, often being staffed by retired military personnel.

New systems and technologies are finding their way into the intelligence community to support the analysis of the vast quantities of information. Three groups of analytic solutions are being distinguished: information fusion, data-mining and visual analytics.[68] Information fusion methods have been developed to automate the process of detection, classification and prediction of phenomena in various fields. Concurrently, data-mining techniques help to discover identical patterns from a multitude of different sources. Visual analytics are also being used to link and analyse large amounts of data stored in many different databases.[69] As has been shown, there are a multitude of challenges to overcome. Concurrently, there are many exciting and innovative developments which provide ample opportunities for analysts.

Conclusion

The nature of complex emergencies is multifaceted, especially since armed conflict is involved. Humanitarian and peace operations are undertaken in challenging environments, especially for those involved in the crucial field of information-gathering. This chapter has argued that a single-faceted understanding of any humanitarian situation is insufficient. Military and civilian leaders in POs need to gain a comprehensive understanding of the conflict area. Currently used intelligence models mainly focus on military aspects and are often focused on instruments of national power. They are not fit for purpose to build a comprehensive understanding of the situation from the point of view of the local populace. The proposed HSI model, however, is built around the civil populace. It analyses the diverse threats facing local populations from a large number of different relevant angles, as well as considering the relationships between these threats.

In traditional military operations, OSINT is usually regarded as less important than the directly gathered traditional military intelligence. However, the more permissive environment of peace operations allows for a more active role for OSINT. This paired with the ITC revolution means that OSINT has greater relevance, and POs have more opportunities to harness its power.

For the proposed HSI concept, OSINT can be a major, if not the principle, source of information.

However, the problem with open source information is less the lack of available information and more the overwhelming abundance of it. Looking for timely and accurate information requires efficient searching strategies. Professionals with skills in information-mining can use their training to ask the right questions and use the right search methods, helping to find the best answers. They can evaluate the reliability of sources and use corroboration across sources to verify information.

To analyse the wealth of collected HSI in complex emergencies, military staff need to be augmented by civil experts. Such experts can be drawn from a range of organisations, such as academic bodies, NGOs, governmental organisations and local organisations. However, using such experts does not come without challenges. Reluctance for civil-military cooperation on both sides needs to be overcome, something that should be easier in peacekeeping than in war-fighting. Also, there are a number of difficulties faced by both civilian and military personnel to operationalise joint intelligence efforts in the field.

With the enormous increase in computing power and software, intelligence analysts can expect new solutions to take some of the burden from their shoulders. However, the challenges of organising information-gathering and analysis in hostile environments remain. The value of OSINT, and the need to harness it in HSI, provides a new frontier for information exploration in peace operations. Hopefully this will allow the international community to better deal with complex emergencies in the future.

Notes

1. This chapter is believed to be the first work to propose and elaborate on the concept of HSI. The only published source that we find to use the term HSI was in conjunction with police operations, using domestic UK examples. See James Sheptycki, 'Policing, Intelligence Theory and the New Human Security Paradigm: Some Lessons from the Field', in Peter Gill, Stephen Marrin and Mark Phythian (eds.), *Intelligence Theory: Key Questions and Debates* (New York: Routledge, 2009), p.166.
2. 'Complex/Manmade Hazards: Complex Emergencies', *International Federation of Red Cross and Red Crescent Societies*, http://www.ifrc.org/en/what-we-do/disaster-management/about-disasters/definition-of-hazard/complex-emergencies/.
3. Ibid.
4. NATO uses the term 'peace support operation' (PSO) to describe such operations, while the UN usually uses the more traditional term 'peacekeeping operation' (PKO). The US uses the broader term 'peace operation'. That more encompassing term is also preferred by us.
5. An early and insightful distinction between conventional and emerging threats was offered by General Al Gray, then commandant of the US Marine Corps, in his article 'Global Intelligence Challenges in the 1990's', *American Intelligence*

Journal (Winter 1989–1990), pp.37–41. He characterised the emerging threat as non-governmental, non-conventional, dynamic or random, non-linear in its development of force capabilities, without constraints (rules of engagement), with unknown doctrine, no established indications and no warning network that could be monitored, and an unlimited fifth column unknown to conventional counterintelligence organisations.
6. Brad E. O'Neill, *Insurgency & Terrorism: From Revolution to Apocalypse*, Second edition (Washington, DC: Potomac Books, 2005), p.32.
7. Department of National Defence, B-GJ-005-307/FP-030, *Peace Support Operations* (Ottawa, ON: DND Canada, 2002), pp.5–8.
8. Ibid., p.2.
9. Benjamin Perrin, 'Guns for Hire – with Canadian Taxpayer Dollars', *Human Security Bulletin* (2008), Vol. 6, No. 3, p.5, http://www.redr.org.uk/objects_store/security_privatization_-_challenges_and_opportunities_2008_.pdf.
10. Christopher Spearin, 'What Manley Missed: The Human Security Implications of Private Security in Afghanistan', *Human Security Bulletin* (2008), Vol. 6, No. 3, p.8, http://www.redr.org.uk/objects_store/security_privatization_-_challenges_and_opportunities_2008_.pdf.
11. Ibid.
12. Department of National Defence, B-GJ-005-307/FP-030, *Peace Support Operations*, pp.5–8.
13. Humanitarian Policy Group, *HPG Research Report, Resetting the Rules of Engagement, Trends and Issues in Military-Humanitarian Relations* (London: HPG, 2006), p.22.
14. Department of National Defence, *B-GJ-005-200/FP-000, Joint Intelligence Doctrine* (Ottawa, ON: DND Canada, 2003), pp.1–4.
15. Major Rob Sentse (RNLA), 'Influencing the Human Terrain. Market Your Product' (2010), http://www.scribd.com/doc/168500960/Influence-Behaviour-Market-Your-Product.
16. Major Rob Sentse (RNLA), 'The African Boulevard of Broken Dreams', *American Intelligence Journal* (2012), Vol. 30, No. 1, p.21. http://www.scribd.com/doc/132071791/The-African-Boulevard-of-Broken-Dreams-American-Intelligence-Journal-Volume-30-2012.
17. Ibid.
18. Ideally the COP should be common to all actors in the PO. However, in reality this is not the case, since many organisations still pursue their own goals and sometimes refuse to fully cooperate and share information with other actors. For instance, the International Committee of the Red Cross (ICRC) in many cases abstains from close cooperation with any military force in the AOR, even with the forces conducting the PO since the ICRC does not want to compromise its neutrality. Nowadays the COP is mostly common to just contributing military forces.
19. The AOI is often larger than the AOR, since events in neighboring areas often influence the state of affairs in the AOR.
20. Figure 7.1 is based on information found in Pauline Kerr, 'Human Security', in Alan Collins (ed.), *Contemporary Security Studies* (New York: Oxford University Press, 2006), p.99.
21. The HSI model was developed by Fred Bruls, under the supervision of Dr Dorn, as part of a thesis for the Master in Defence Studies programme at the Canadian Forces College in 2011. The thesis is published at http://www.walterdorn.org/pdf/HumanSecurityIntell-PSO_Bruls_MDS-Paper_ForPublicRelease_Feb2012.pdf. For

the UN report in question, see UNDP, *Human Development Report 1994* (New York: Oxford University Press, 1994), p.24.
22. The craft of intelligence does appear to be evolving more rapidly than in the past. The concept of HSI – an outcome to be sought – should be distinguished from the concept of full-spectrum human intelligence (the sources) as outlined by Robert David Steele in his monograph, *Human Intelligence: All Humans, All Minds, All the Time* (Carlisle, PA: Strategic Studies Institute, 2010).
23. Col. Jack D. Kem, 'Understanding the Operational Environment: The Expansion of DIME', *Military Intelligence* (2007), Vol. 33, No. 2, p.1.
24. Acronyms: DIME (diplomatic, information, military and economic); DIMEFIL (diplomatic, information, military, economic, financial, intelligence and law enforcement); ASCOPR (areas, structures, capabilities, organizations, people and events); and PMESII (political, military, economic, social, information and infrastructure).
25. Personal conversation between one of the authors (Bruls) and LTC Martien Hagoort, staff officer at the Netherlands Military Intelligence School, 17 August 2011. LTC Hagoort explained that DIME and DIMEFIL are not being used on the operational and tactical levels of POs. The only models to be found today are ASCOPE and PMESII. Therefore the Netherlands Military Intelligence School does not train its students in models such as DIME or DIMEFIL. Consequently these models are not discussed in this chapter.
26. United States Center for Army Lessons Learned, *Handbook 10–41: Assessment and Measures of Effectiveness in Stability Ops: Tactics, Techniques and Procedures* (Fort Leavenworth: Combined Arms Center (CAC), 2010), p.6.
27. Explanation of dimensions paraphrased from Kem. 'Understanding the Operational Environment: The Expansion of DIME', p.4. The article gives an explanation of both old and recent models, including DIME, DIMEFIL, MIDLIFE, PMESII and ASCOPE.
28. See US Army, *Field Manual 3-0: Operations* (Washington, DC, 2008).
29. PMESII was originally meant to analyse the instruments of power that an opposing party possessed and which of these instruments this party was likely to use. This analysis was combined with an estimate of the optimum mix of instruments to be used to coerce this party to comply with the demands of the host nation where the analysis was made.
30. Personal conversation between one of the authors (Bruls) and LTC Martien Hagoort on 17 August 2011. L. T. C. Hagoort explained that PMESII is the only model that is being taught at the Netherlands Military Intelligence School. However, he opposes overreliance on a single intelligence model, given the risk of tunnel vision. He therefore would like to see ASCOPE return to the syllabus.
31. United States Center for Army Lessons Learned, *Handbook 10–41: Assessment and Measures of Effectiveness in Stability Ops*, p.6.
32. Kem, 'Understanding the Operational Environment', p.6.
33. Ibid.
34. Maj. Brian M. Ducote, *Challenging the Application of PMESII-PT in a Complex Environment* (Fort Leavenworth: United States Army Command and General Staff College, 2010), p.10.
35. US Army, *Field Manual 3-0: Operations*, pp.1–5; Ducote, *Challenging the Application of PMESII-PT in a Complex Environment*, p.7.
36. Ducote, *Challenging the Application of PMESII-PT in a Complex Environment*, p.20.
37. Ibid., p.20, p.53, p.38.

38. Personal observation by one of the authors (Bruls) in the position of Information Manager of the G2 branch within Task Force Uruzgan from August 2009 to February 2010.
39. Ducote, *Challenging the Application of PMESII-PT in a Complex Environment*, p.39.
40. Montgomery McFate, 'The Military Utility of Understanding Adversary Culture', *Joint Forces Quarterly* (2005), No. 38, p.44. Although cultural could be considered to be part of the 'social' dimension of the PMESII(-PT) model, McFate argues that cultural knowledge is often lacking. Culture is certainly not sufficiently taken into account in the application of the PMESII(-PT) model.
41. The term 'human security' is a debated one. The narrow definition is freedom of individual human beings from physical threats. The broader definition includes all manner of threats (for example, to health, food sources, finances, stability and identity). The masters thesis mentioned in note 21 contains a description of the most important approaches to define the term. See http://www.walterdorn.org/pdf/HumanSecurityIntell-PSO_Bruls_MDS-Paper_ForPublicRelease_Feb2012.pdf.
42. UNDP, *Human Development Report 1994* (New York: Oxford University Press, 1994), p.24.
43. Ibid., p.30.
44. Ibid., pp.31–32.
45. Ibid., pp.32–33.
46. It is tempting to designate this model with an acronym referring to the factors that it covers, as do the other intelligence models discussed in this chapter (such as ASCOPE and PMESSI). However, avoiding an acronym has benefits. As shown above, intelligence models in today's operations are often used as a simple exercise of quantifiable fact-finding to feed into different dimensions, while the links between the dimensions are often ignored, leading to information with limited value.
47. Ducote, *Challenging the Application of PMESII-PT in a Complex Environment*, p.53.
48. A. Walter Dorn, 'Intelligence-led Peacekeeping: The United Nations Stabilization Mission in Haiti (MINUSTAH), 2006–07', *Intelligence and National Security* (2009), Vol. 24, No. 6, pp.805–835.
49. Anthony Olcott, *Open Source Intelligence in a Networked World* (New York: The Continuum International Publishing Group, 2012), p.109.
50. Nada J. Pavlovic, Lisa Casagranda Hoshino, David R. Mandel and A. Walter Dorn, *Indicators and Indices of Conflict and Security: A Review and Classification of Open-Source Data*, Technical Report, Defence Research and Development Canada, Toronto, September 2008, http://oai.dtic.mil/oai/oai?verb=getRecord&metadataPrefix=html&identifier=ADA494833, p.3.
51. Human Security Report Project, *Human Security Report 2009/2010: The Causes of Peace and the Shrinking Cost of War* (New York/Oxford: Oxford University Press, 2011), p.126.
52. Ibid., p.124.
53. In many military operations, as in Afghanistan and Iraq, connections were established over narrowband satellite connections. Access to the Internet was very slow, making the search for information from open sources an extremely time-consuming exercise.
54. Arpad Palfy, 'Intelligence Information Management in Joint Environments', *Vanguard*, December 2010/January 2011, http://vanguardcanada.com/intelligence-information-management-in-joint-environments/

55. In order to adopt information management as a topic in the curriculum of intelligence personnel, in 2010 a working group was established at the Netherlands Military Intelligence School in which one of the authors (Bruls) plays an advisory role.
56. Personal observation in the position of Information Manager of the G2 branch within Task Force Uruzgan from August 2009 to February 2010.
57. Personal conversations with the MRTF S2 and Australian liaison officers within the ASIC within the G2 branch within Task Force Uruzgan over the period August 2009 to February 2010.
58. Ibid.
59. MGen Michael T. Flynn, Capt. Matt Pottinger and Paul D. Batchelor, *Fixing Intel: A Blueprint for Making Intelligence Relevant in Afghanistan* (Washington, DC: Center for a New American Society, January 2010), http://www.cnas.org/files/documents/publications/AfghanIntel_Flynn_Jan2010_code507_voices.pdf, p.9.
60. See Adam B. Siegel, 'Intelligence Challenges of Civil-Military Operation', *Military Review* (September/October 2001), pp.45–52.
61. Robert David Steele, 'Information Peacekeeping and the Future of Intelligence: The United Nations, Smart Mobs and the Seven Tribes', in Ben de Jong, Wies Platje and Robert David Steele (eds.), *Peacekeeping Intelligence: Emerging Concepts for the Future* (Oakton, VA: OSS International Press, 2003), pp.201–255.
62. ibid.
63. Royal Netherlands Army, *Command Support in Land Operations: Doctrine Publication 3.2.2.1. Study Draft 4 (CONCEPT)*. pp.6–19. Special thanks to LTC Chris Rump, RNLA for providing us with this draft.
64. McKinsey, *Big Data, Small Wars, Local Insights*, http://voices.mckinseyonsociety.com/big-data-small-wars-local-insights-designing-for-development-with-conflict-affected-communities/ (accessed 1 May 2013).
65. See Siegel, 'Intelligence Challenges of Civil-Military Operation'.
66. For an overview, see Victoria Wheeler and Adele Harmer (eds.), *Resetting the Rules of Engagement, Trends and Issues in Military-Humanitarian Relations* (London: Humanitarian Policy Group, March 2006), http://www.odi.org.uk/sites/odi.org.uk/files/odi-assets/publications-opinion-files/273.pdf, p.68.
67. Personal observation by one of the authors (Bruls) in the position of Information Manager of the G2 branch within Task Force Uruzgan from August 2009 to February 2010.
68. Gregor Pavlin, Thomas Quillinan, Franck Mignet and Patrick de Oude, 'Exploiting Intelligence for National Security', in Babak Akhgar and Simeon Yates (eds.), *Strategic Intelligence Management* (Waltham, MA: Butterworth-Heinemann, 2013), p.167.
69. Ibid., p.186.

Part IV

Open Source Intelligence and Counterterrorism

8
Detecting Events from Twitter: Situational Awareness in the Age of Social Media

Simon Wibberley and Carl Miller

Introduction: Twitter's CNN moment

'Helicopter hovering over Abbottabad at 1am (is a rare event)', tweeted Sohaib Athar on 2 May 2011. Other tweets quickly followed: the helicopters (he quickly realised there were more than one) were non-Pakistani, there was a window-shaking explosion, a 'gun fight', a crash and an army cordon.[1] Athar was live tweeting an operation that had been planned and executed in the darkest depths of secrecy: the US SEAL raid on the home of Osama bin Laden. It was Twitter's CNN moment – the emergence of a new and significant channel for people to report on, and learn about, important events.[2]

Since the arrival of the Foreign Broadcast Information Services and their allied institutions over 70 years ago, 'open' media sources – from clippings of *Krasnaya* to Beijing *Xinhua* news – have been an important way to learn about significant 'events'; political, cultural, commercial or emergency. Indeed, whether urban disorder,[3] an epidemic[4] or an underground nuclear weapons test,[5] rapidly identifying that something has happened and dependably supplying situational awareness – the 'what', 'where', 'whom' and 'when' – has been and remains a fundamental contribution of OSINT.

Maintaining and enhancing situational awareness is vital to counterterrorism. It is vital to stop terrorist attacks from happening and to protect society against them when they do occur; to prevent people from becoming terrorists and to effectively pursue them when they do. It underlies everything from the effective response of emergency services to an attack, to noticing the radicalising influence of a trial, to retasking resources to respond to shifts in the international threat environment. In this context, our ability to discern situational awareness in a changing world is a key part of the challenge of ensuring that our response to terrorism is effective.

The explosion of a new, 'social' media – those platforms, Internet sites, apps, blogs and fora that allow users to generate and share content themselves – have drastically changed the type and number of open sources that can now be exploited to maintain situational awareness. Our Facebook 'likes', Instagram posts, Twitter tweets, FourSquare check-ins and Pinterest pins together constitute ever-increasing proportions of our cultural and intellectual activity. Taken together, social media is the largest reservoir of information about people and society that we have ever had: huge, naturalistic and constantly refreshing bodies of behavioural evidence that are, in digital form, inherently amenable to collection and analysis.[6]

The emergence of Twitter specifically has huge consequences for our ability to quickly detect events that have occurred, and find out information about them from open sources. People use Twitter to microblog no more than a few sentences, a few individual images or links at a time. Twitter's 200 million active users together post 400 million of these microblogs – tweets – daily. The 'tweet-stream', a gushing torrent of constantly arriving tweets, is a new layer of information in society. It is a chaotic and diverse deluge of digital commentary, arguments, discussions, questions and answers.

Offline events cast ripples into this tweet-stream. Twitter users – like Sohaib Athar – report on events first-hand. They use it as a new form of digital 'citizen witnessing', a type of first-person reportage in which ordinary individuals temporarily adopt the role of a correspondent journalist in order to participate in news-making, often spontaneously when they find themselves at the scene of an important event, crisis or disaster.[7] Others on Twitter magnify these initial ripples – requesting more information about them, 'brokering' information between parties, checking information and adding additional information from other sources and propagating information that already exists within the social media stream.

These ripples together form 'twitcidents' – clouds of online reaction that shadow events that have occurred offline.[8] They are a collectively authored digital annotation of the event, containing questions, interpretations, condemnations, jokes, rumours and insults. Although now still novel, twitcidents will become a routine aftermath, a usual way in which society reacts to and annotates the events that it experiences. Indeed, twitcidents will become an important dimension of the events themselves.

This chapter is about harnessing the tweet-stream to detect an event's online shadow, and therefore the event itself. It presents two new methods by which to do this, and two case-studies of their use. It ends with a discussion about what these case-studies suggest about the role of Twitter event-detection technology to maintain situational awareness in the social media age, particularly in the context of counterterrorism, and the ethics involved in doing so.

Finding events in the 'tweet-stream': The A, B, C of automatic event detection

Detecting events means finding those online ripples that they cast into the tweet-stream. To this end, the scale and velocity of Twitter – like many social media platforms – is a double-edged sword. The tweet-stream arguably provides both more 'signals' of events and more 'noise' – irrelevant, wrong and totally useless tweets – than ever before.[9] Drawing a clear and dependable picture of an event from this often overwhelming mass of information of greatly ranging value, especially at the speed demanded by the tempo of the event itself, is a formidable intellectual and technological challenge.

However, the explosion of this kind of data has been accompanied by the growth of a suite of primarily statistical and computational 'big data analytics' tools. These are a collected body of computational, statistical and machine-learning techniques that are capable of making sense of data of great complexity and scale, and with great speed.

This chapter presents two new methodological systems that leverage these new technologies to detect events. Both are automated, computational and algorithmic attempts to analyse (a) an input to detect (b) 'events' with (c) targeting. These concepts are elaborated upon below:

Input. Event-detection algorithms need data to work on. Both methods presented here collect a portion of the tweet-stream automatically by connecting to the platform's application programming interface (API). The API is a portal that acts as a technical gatekeeper of the data held by the social media platform. There are three on Twitter, and while they differ in important ways – such as the type of data they allow researchers to access, and the quantities that they produce it in – they all allow the researcher to download tweets in large quantities and structured formats. They also produce with each tweet a body of metadata, such as who produced the tweet and when.[10]

Event definition. Conceptually, for algorithms to detect events, there needs to be a definition of an 'event'. We take this definition in its broadest and simplest form: 'the fact of anything happening'.[11] To be detected, events need only to have a 'when', the time at which the event happened, or started happening. Finding additional attributes of an event – including 'where', 'who' and 'what' – relates to its 'characterisation', and also something that both methods will attempt.

Targeting. The detection of events from the tweet-stream can be either targeted or untargeted. Targeted event detection tries to find particular kinds of event that we know something about: the kind of event that will happen, when an event will happen or the people who will make it happen. However, we do not always know what kind of event will be important, and untargeted event detection – a more challenging task – makes no judgement about the kind of event to be detected *a priori*.

The two methods presented are united on points (a) and (b), but crucially differ on point (c). A different case study is used to illustrate each method.

Method 1: Targeted event detection: London 2012

During the 2012 London Summer Olympics, much was known about the kind of events that would occur during the course of the event. We knew that a defined body of people (Olympians) were to compete in a set number of events for a limited body of prizes (the coveted Olympic medals). Working on this general body of certainties, targeted event detection was used to detect when a medal had been won, and find further details of its winning, such as who the winner was and what the medal was for.

Step 1: Collect tweets

The first step involved the collection of tweets from the tweet-stream to be analysed. Because the event we were trying to detect was defined, the information we could collect could also be (loosely) defined. Accessing Twitter's APIs, 30,470,932 tweets posted between 18 July and 13 August 2012 were collected. These tweets all contained at least either the first or the last name of an Olympian competing in the games. They were stored in a dedicated Structured Query Language (SQL) database.

Step 2: Measuring tweets over time

Each 'medal win' during London 2012 was a media and social event, provoking reaction on various fora, including Twitter. The ripples caused by each win spread through the tweet-stream in the form of a surge in related chatter. These surges were therefore good indicators of the events that caused them.[12] To detect surges in the tweet-stream, the 30,470,932 gathered tweets were plotted over time. This provided a very general sense of the overall surge of tweets flowing through Twitter that, through virtue of possibly mentioning an Olympian, were roughly associated with the topic of the Olympics.

Figure 8.1 shows a window of this timeline, counting the number of tweets appearing between roughly 17:00 and 19:30 on 31 July 2012. Towards the end of this time window, marked in vertical line marker, the rate of tweets sharply increases.

Step 3: Measuring change in the rate of tweets over time

Next, the scale of each surge was measured. To do this, first the number of tweets arriving in a specific time segment (in this case, two minutes) was calculated.[13] The change in the rate of tweets was then calculated by comparing

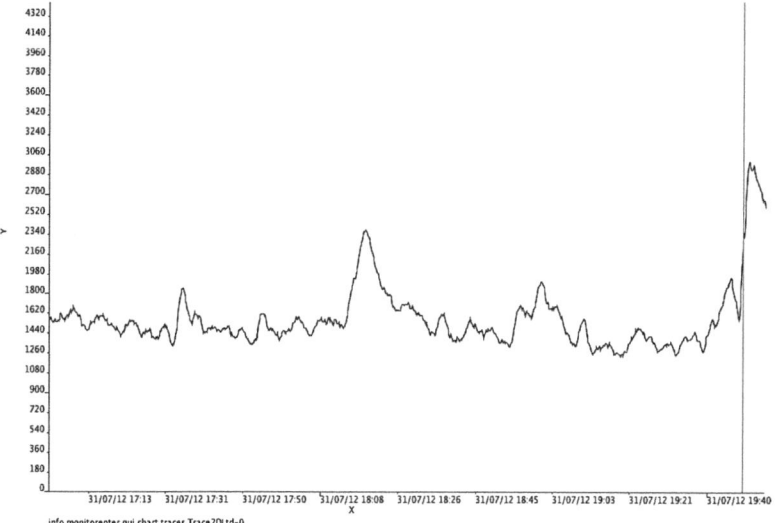

Figure 8.1 Tweets containing either the first or the last name of an Olympian arriving in the tweet-stream every two minutes

the number of tweets arriving in every time segment against the number of tweets arriving in the previous equal time segment. This produced a value for every time segment as a ratio of the previous time segment.[14]

Figure 8.2 therefore shows the same tweets sent between 17:00 and 19:30 on 31 July 2012, but where the tweets are grouped into time segments of two minutes and where each two-minute segment is expressed as a ratio of the size of the previous one.

The spike shown in Figure 8.1 is now shown to be a two-minute time segment that contains 1.84 times the tweets in the two-minute segment before it. This passes a threshold of 1.7 that was inductively set when observing, externally to this case-study, the behaviour of the tweet-stream during other London 2012 medal wins.[15] Surges surpassing 1.7 were found to be a good indicator of a surge significant enough to be the result of a medal win.

Therefore, at the end of step 3, the identification of a surge in the tweet-stream suggested the time of a candidate event. Shown in Figure 8.3, pale grey marks the 'pre-event' tweet-stream, and dark grey a surging tweet-stream possibly indicating an event. This, the blue section of the graph, is denoted as a possible 'event zone': the immediate online aftermath of an event.

Step 4: Characterising the 'event'

Thus far we know nothing about this possible event. We do not know if it is indeed a medal win, and if so who won what.

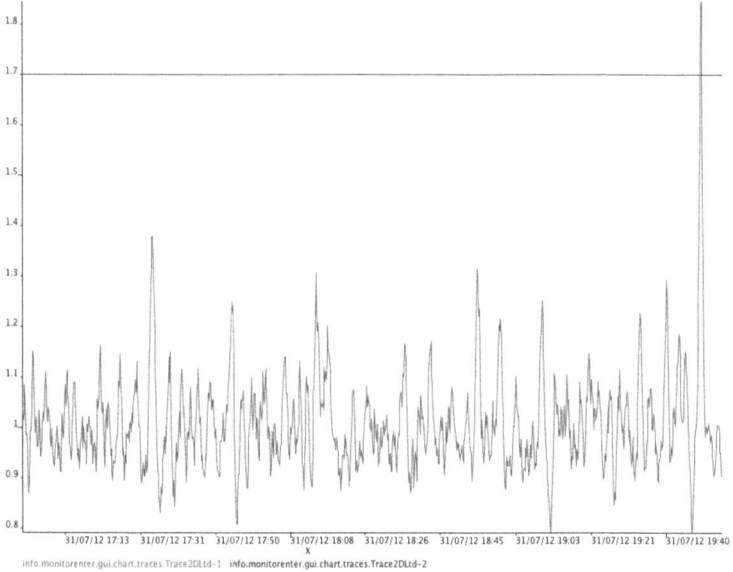

Figure 8.2 Number of tweets sent every two minutes, between 17:00 and 19:30 on 31 July 2012, expressed as a ratio of the number sent in the previous two minutes

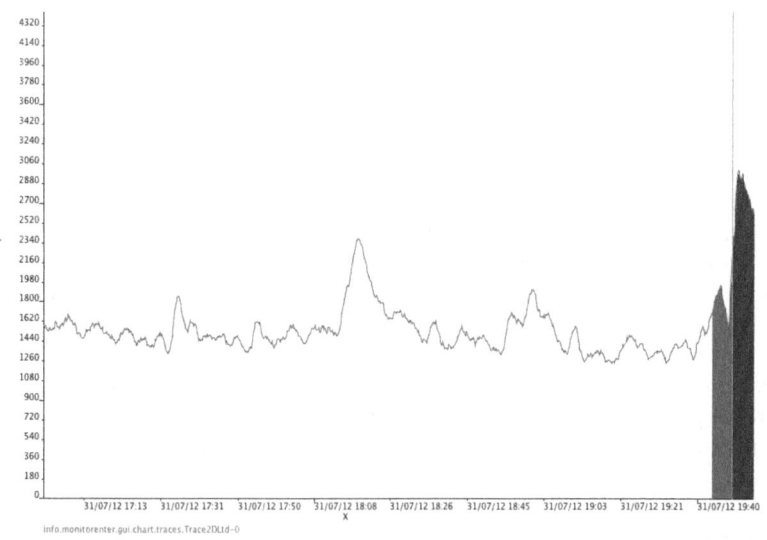

Figure 8.3 A candidate event: A 'pre-event' and possible 'event' tweet-streams

To learn additional facts about this candidate event – to characterise it – we need to look underneath the numbers and at the actual content of the tweets themselves. Here we face a 'wheat and chaff' problem. Taken from a loosely targeted tweet-stream of tens of millions, the 'event zone' is composed of 2313 tweets on lots of different topics; many about the event in question, but many others talking about other events, or no events at all.

To characterise the event, we needed to find what was special and therefore discriminative about the tweets in the event zone compared with the tweets in the general timeline. To do this, every individual term in every tweet in the event zone was identified. The chance of a term occurring in any tweet in the event zone, and the chance of it occurring in any tweet in the general tweet-stream, were measured.[16] Then the difference between these two probabilities was measured, and the term was given this measure as a score.[17] The higher the score, the more likely the term was to be present in event-zone tweets than in those in the general tweet-stream.

Every tweet in the event zone was then scored as the sum of the scores of each of the terms that it contained. This produced a list of the top-scoring tweets that, by this system of reckoning, were the ones that were most particular and discriminative of the event zone compared with the general tweet-stream, and therefore most likely to be about the event.

For the event in question here, of the 2313 tweets in the event zone, Table 8.1 shows the highest rated and lowest rated tweets.

Table 8.1 Highest rated and lowest rated tweets

Tweet text	Score
Gold Gold Chad le Clos by the fingertip #teamSA #london2012 Gold Gold Chad le Clos by the fingertrip #teamSA ... http://t.co/EDtYwkbu	0.715
20th of a second between Chad Le Clos and Michael Phelps in the 200M Butterfly!? Wow, what a final! Credit to Le Clos! #London2012Olympics	0.618
Here comes Michael Phelps – WOW! Misses gold by 0.01 seconds! Phelps takes silver. South African Chad Le Clos wins gold #London2012 #Olympx	0.570
Wow Michael Phelps misses gold by 0.01 seconds! Phelps takes silver. South African Chad Le Clos wins gold #London2012 #TeamUSA	0.595
Wow what a race! Chad Le Clos and Michael Phelps 200m ... Guess who won? Le Clos! Representing #RSA! These olympics aren't as boring anymore.	0.587
How awesome was that??? Chad Le Clos just beat Michael Phelps!!! Gold for Chad, Gold for SA! #london2012	0.552

Table 8.1 (Continued)

Tweet text	Score
Phelps gets a silver by .05 seconds, Chad le Clos wins the gold in Phelps' favorite event, the 200M butterfly. #London2012 #Olympics	0.551
What. A. Race. Phelps Vs Le Clos. Woahhhhhh. What a finish from Le Clos... Stelllaaaaarrrr bru #London2012 #Swimming	0.541
Chad Le Clos edges Michael Phelps for 200 fly gold, Phelps ties record for most career Olympic medals. #London2012	0.534
Chad Le Clos edges Michael Phelps for 200 fly gold, Phelps ties record for most career Olympic medals. #london2012... http://t.co/Ds1Joyb3	0.534
...	...
When I was 10 I dreamed of going to the Olympics, the furthest I got was European Champion at the age of 12 and then I stopped...	−0.066
took gymnastics from 18 ms – 10 yrs then quit cuz I didn't think I was good I find out I prob wld hv been in the Olympics fml	−0.067
@raytetreault: I'm gonna get the Olympics ring tattoo and just tell everyone I was in the Olympics. #soundsgood	−0.067
Sorry for the Olympics spamming but I've loved the Olympics since I ripped my ankle ligaments in 2004 and watched it 24/7 :)	−0.067
@johnebeck I kept it vague. dont' want to be guilty of ruining the olympics for anybody. it's olympics related. do you wanna know?	−0.067
I'm so sick of the olympics. I just want to watch my damn shows but I can't cause the olympics is on every damn tv in my house :P	−0.080
When guys talk about the Olympics, they actually talk about the Olympics. IF girls talk about the Olympics, they talk about the hot guys.	−0.087
Dafuq?! I just barely noticed that I was following the olympics people I don't even like the Olympics	−0.087
Meet the Best Olympics Advertisers Who Can't Say 'Olympics': You don't need to be an official Olympics 2012... http://t.co/lHN0MQpq #MAFA	−0.087
Meet the Best Olympics Advertisers Who Can't Say 'Olympics': You don't need to be an official Olympics 2012 spon... http://t.co/IXwzFHDk	−0.087

These high-scoring tweets share a handful of common terms that are highly discriminative of the event zone in comparison with the general tweet-stream. They commonly mention two Olympians – Chad Le Clos and Michael Phelps – and a swimming event, the 200 m butterfly or 'fly' gold medal race.

Overall, this case of targeted event detection painted a clear picture. At around 19:40 on 31 July 2012, a ripple spread across the tweet-stream. A surge of tweets generally mentioning Olympians occurred, and this surge was significant enough in scale to pass a threshold that was set by looking at other medal wins during the 2012 Olympics. The chatter that best characterised this surge focused on the 200 m butterfly gold medal race, and the surprise victory of Chad Le Clos over Michael Phelps. This was accurate. A South African swimmer, Chad Le Clos, did indeed beat Michael Phelps by five one-hundredths of a second on the evening of 31 July 2012.[18]

Method 2: Untargeted event detection: The August 2011 riots

It is sometimes unforeseen events that are more important to detect quickly and accurately. On Thursday 4 August 2011, Mark Duggan was shot and killed by a police officer in North London. The resulting violence spread throughout Tottenham, Tottenham Hale and Wood Green on 6 August, then more widely across London on 7 and 8 August, and finally to other cities by 9 August. This wave of violence, looting and anger, the most significant urban unrest seen in decades, cost five lives and hundreds of millions of pounds in damaged property. It took many, including the police, by surprise.

Yet as early as the morning of 6 August, social media channels showed burgeoning hostility, peppered with explicit threats against the police. From 7 August, social media information indicated the possible spread of disorder to other parts of London, then other cities in England. Over the next few days, messages indicating criminal intent or action ratcheted in huge numbers across a number of social media platforms, including Twitter.[19] Following the riots, the police acknowledged that they had missed these warning signs. Her Majesty's Chief Inspector of Constabulary noted: 'With some notable individual exceptions, the power of this kind of media (both for sending out and receiving information) is not well understood and less well managed.'[20]

Method 2 therefore confronts the more challenging problem of detecting events using Twitter that, like the riots, were unexpected. Without prior knowledge of the type, nature, scale or significance of events that should be detected, there are many things that untargeted event detection cannot do. For example, it cannot target or gather any particular part of the tweet-stream as being particularly relevant to the event (as was the case in method 1). Nor can it make any assumptions regarding the timescale over which the event may occur, the kind of people who are likely to be talking about it or what they are likely to be talking about.

Bearing these difficulties in mind, the tweet-stream of 6 August 2011 will be analysed. Without drawing on any retrospective knowledge of what was to occur, a method will be presented to isolate a tell-tale clustered burst of tweets indicating the occurrence of events associated with rising tensions in Tottenham.

Step 1: Collect tweets

Method 2 began with a very broad, almost random (and importantly not targeted to be riot-related) swathe of the tweet-stream. Some 70,000 tweets were collected, all of which were created between 18:00 and 24:00 on 6 August 2011. These tweets were selected both because they fell within this time window and because the authors had identified themselves to Twitter as being in the UK.

Step 2: Measure the tweet-stream's rate over time

These 70,000 tweets were plotted over time in the same way as in method 1. This time, however, we were unable to measure the rate of any predefined set of words due to the untargeted nature of the collection. As Figure 8.4 shows, the general tweet-stream overall remained fairly flat on the evening of 6 August. It rose slightly (as is usual) when people returned from work.

Step 3: Calculate each term's signal

Whilst the tweet-stream, taken as a whole, was not behaving anomalously on the evening of 6 August, step 3 sought to determine whether distinct parts of it were. The underlying principle of untargeted event detection is to analyse the whole of the tweet-stream (or a random sample of it), without prejudice, in order to find sections that are behaving in unusual ways that may reflect online chatter about an offline event.

The specific behaviour that this method tries to spot within the tweet-stream is 'clustered bursts' or groups of words in the tweet-stream that all suddenly become used more in a short timeframe. The assumption is that some events provoke 'clustered bursts' – unusual eddies of language use within the tweet-stream.[21]

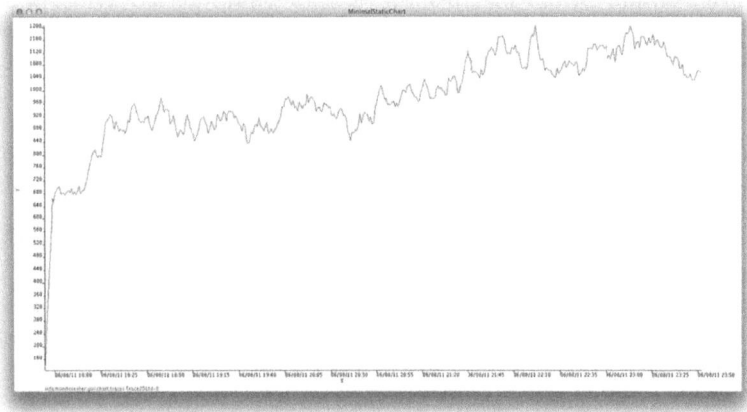

Figure 8.4 Tweet-stream on 6 August 2011 between 18:00 and 24:00

Stage 1: Signal measuring 'Burstiness'

To find these indicators, the method treats every term that occurs in the Twitter stream as each forming its own wave, or 'signal'.[22] This wave represents each term's change in usage (in the tweet-stream) over time. Terms that consistently occur in the tweet-stream – at any quantity as long as the quantity is unchanging – have a flat wave. Terms that suddenly 'burst' suddenly become much more frequently used than they previously were – they have a rippled or spiked wave. To calculate each term's signal, a term's short-term usage (the frequency of its appearances in tweets posted in one time interval, in this case five minutes) is compared with a backdrop of the tweets' long-term usage (the frequency of its appearances in a historical floating window).[23]

For illustration, Figure 8.5 shows the stage 1 signals of some terms contained within the sample of 70,000 tweets. As is clear, compared with 'lol', words like 'Tottenham', 'riot' and 'police' (which the untargeted method has not analytically isolated) are beginning to burst.

Stage 2: Cleaning the signal and measuring over different scales

As Figure 8.5 illustrates, stage 1 signals are jagged and noisy. To be compared with each other, they need to be 'cleaned'. To do this a signal-processing technique called 'wavelet analysis' is used.[24] Wavelet analysis does this by fitting (or convolving) a 'wavelet' signal to the original stage 1 signal over multiple successive time intervals. Each convolution results in an

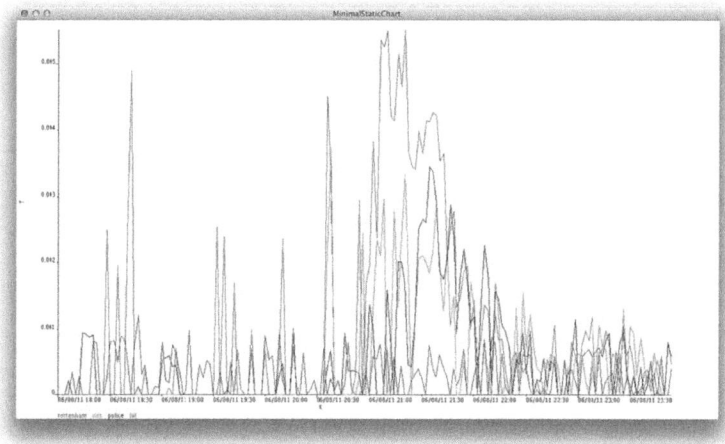

Figure 8.5 Stage 1 signals of some example terms: 'tottenham', 'riot', 'police' and 'lol'

'approximation' of the original signal (the signal with some 'detail', or short-term variation removed), and the 'detail' which is the local residual variation between successive signal approximations.

This process can then be recursively applied to the signal approximation. This allows us to look at the 'burstiness' of a term over different timescales and volume scales simultaneously. Events can happen at different scales – seconds, minutes, hours and even longer – prompting bursts of terms to happen on different scales. When nothing is known about the event being detected, it is important to avoid presuppositions regarding the timescale of the events' occurrence. Analysing the emergence of 'clustered bursts' over different timescales, as shown in Figures 8.6a–8.6e, allows for the detection of events over different scales.

The following series of plots illustrate the process of wavelet analysis on the signal of an example term. Figure 8.6a depicts the original signal and 8.6b–e show the signal approximation in dark grey, and the removed detail in pale grey.

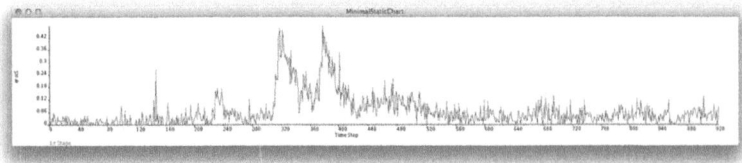

Figure 8.6a   Original signal

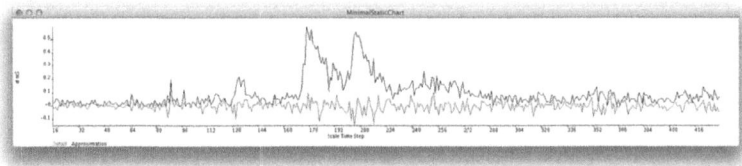

Figure 8.6b   First-level analysis

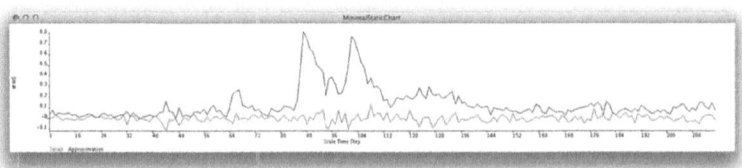

Figure 8.6c   Second-level analysis

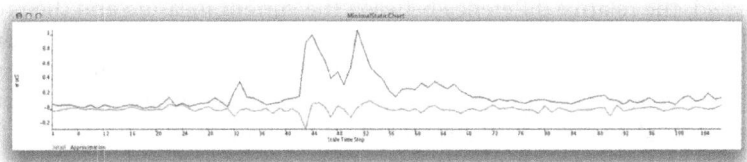

Figure 8.6d   Third-level analysis

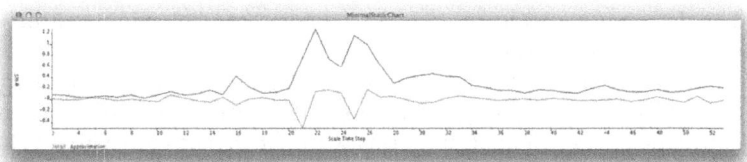

Figure 8.6e   Fourth-level analysis

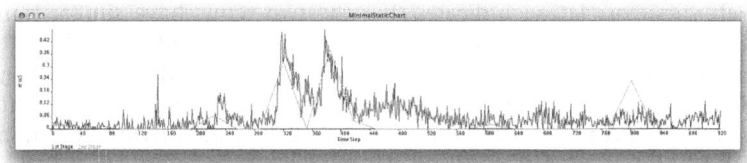

Figure 8.6f   Resulting wavelet analysis

The signal clearly bursts on two distinct occasions. While the bursts are intuitively obvious to the human eye, the small-scale variations in the original signal effectively obscure them from automatic analysis. This remains true for the first two levels of wavelet analysis – the noise in the signal approximation and detail is significant enough to make burst recognition problematic. It is only when you remove detail at the appropriate scale to the event itself (Figure 8.6e) that the detail becomes significant and echoes the spikes that can be tracked by eye.

The final output (Figure 8.6f) of the wavelet analysis component uses the signal variability (the 'detail') of all the scales to calculate whether there is a genuine burst of activity. First it normalises the signal detail measurements with the 'total absolute' detail expressed across all scales.[25] It then takes the normalised entropy over those normalised details. This final calculation forms the measure which the algorithm uses as a cue for activity in a signal. A drop in the normalised entropy from one time interval to the next indicates significant burstiness on at least one of the scales. If a drop is detected, the relative difference between the current measure and the previous one forms the output of the analysis for that time step. In other words, the algorithm continues to analyse the tweet-stream at different scales until there is a significant difference between two different scales: this difference indicates that an event may have occurred.

Step 4: Clustering similar 'cleaned up' signals

The purpose of both calculating and cleaning up the signal of each term contained in our 70,000 tweets was to find those 'clustered bursts' of terms that, in forming a pattern of similarly unexpected behaviour within the tweet-stream, may indicate an event.

To find these clustered bursts, a number of points in each term's stage 2 signal are taken to form the 'burstiness' profile for each term. Before proceeding to the next stage of analysis, terms with flat signals – ones that behave consistently throughout the entire time window and have low 'burstiness' – are filtered out. Then two calculations are carried out. The first compares the signal of every remaining term for similarity with the signal of every other term. This calculation produces each pair of terms' 'cross-correlation': a high cross-correlation indicates that the two terms have similar signals and therefore have burst in a similar way.[26] The second calculation then attempts, on the basis of each term's cross-correlation with every other, to find clusters of terms that all have similar signals (and therefore high cross-correlation scores) to each other.[27] This is called identifying community structure.

The result of these two calculations can be produced as a weighted graph. Figure 8.7 represents the 70,000 tweets collected in step 1, where each term contained within it is a node connected to every other term with a weight corresponding to their similarity, and then segmented into smaller clustered groups of co-similarity.

Step 5: Interpretation of clusters

Figure 8.7 therefore shows many different clusters of terms that may be indications and descriptions of events. The final step in this method is to interpret this graph to find the specific clusters that do in fact indicate events.

In general, small and tight clusters of words are more likely to indicate events than large, loose clusters. A series of thresholds can therefore be applied to Figure 8.7 to discount large loose clusters, and to produce a series of small, tight clusters for consideration.[28]

One of the small, tight clusters buried within the complexity of Figure 8.7 is drawn out and shown in Figure 8.8.

This untargeted method of event detection began with a body of general tweets from the tweet-stream on the evening of 6 August 2011. While the overall volume of the tweet-stream remained constant that evening, under

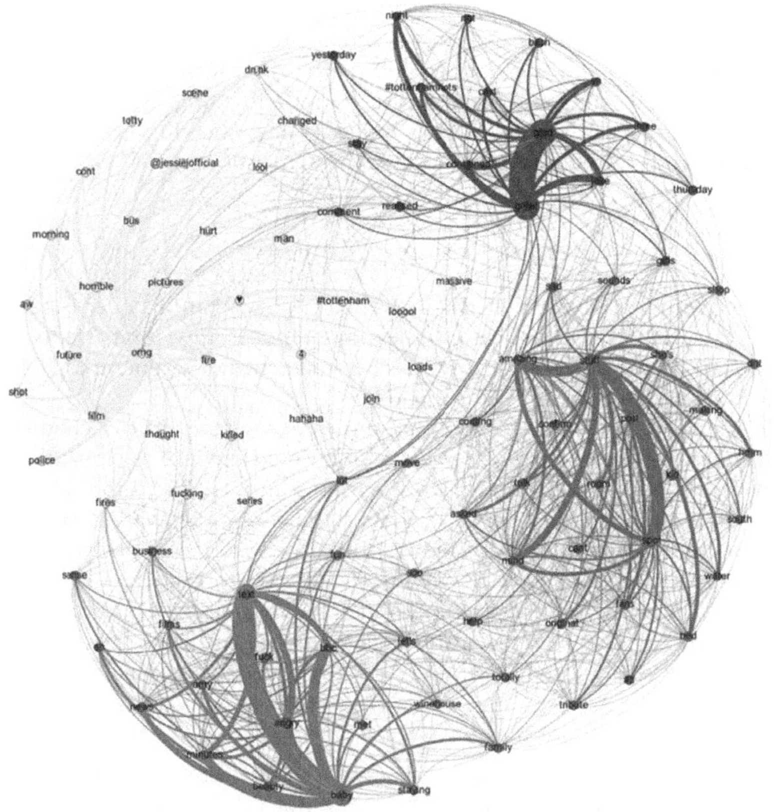

Figure 8.7 Weighted graph showing 'community structure' – similar signals – between terms

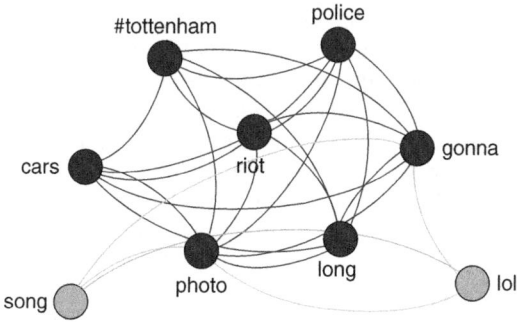

Figure 8.8 Example of clusters drawn out from Figure 8.7

the surface ripples were beginning to form. Clusters of words were suddenly and unusually being used more, and one of these ripples (Figure 8.8) – '#tottenham', 'police', 'cars', 'riots' and (perhaps most ominously) 'gonna', pointed towards the approaching crisis. Only hardening into true significance in retrospect, the tweet-stream was in fact showing echoes of the rising tension in Tottenham that resulted from the shooting of Mark Duggan that day.

Applications of event-detection technology

Neither winning Olympic medals nor engaging in urban disorder is acts of terrorism. However, these case-studies, by demonstrating that events like these can be detected, point towards the emergence of a generic capability that is applicable to countering terrorism.

While the two event-detection methods were explained in this chapter as a number of discrete stages, they both function in practice as continuously and fluidly operating systems. These systems are scalable, duplicable and adaptable, and they can be put to a wider number of uses in varied scenarios. In principle, they can work to reduce the surprise of both anticipated and unanticipated 'hard' and 'soft' events related to terrorism. Hard events are those that physically and discretely happen – a fire, a riot or a terrorist attack. Soft events are ideational and attitudinal shifts that, while more amorphous and less discrete, nonetheless represent a real and marked change to the *status quo*.

Potential applications may include:

- *Situational Awareness of rapidly emerging emergency events.* The event-detection technology can quickly spot the emergence and evolution of events such as riots and terrorist attacks. In these emergency scenarios, traditional communication mechanisms, such as mobile-phone networks, can quickly become saturated. Most recently, during the murder of Lee Rigby in South London and the subsequent English Defence League demonstration, 20,000 tweets flooded into the Metropolitan Police's account, and hundreds of thousands of tweets otherwise relevant to detecting what was happening were also posted.[29]
- *Rapid changes in a group's behaviour.* Sudden and dramatic shifts in a group's dynamics may occur, including angrier, more action-oriented words, increasing dissent or the emergence of breakaway factions. Pathways towards radicalisation often conceive of individuals or groups moving through a number of distinct steps – such as the model New York Police Department: 'pre-radicalization', 'self-identification', 'indoctrination' and 'jihadization'.[30]
- *Rapid change of national narratives.* These might include bursts of nationalist, revanchist or irredentist sentiment that leads to increased political pressure on leaders. This can take the form of calls on leaders to develop

nuclear weapons, or in crisis-situations of taking a 'hard line' against a national adversary.
- *Geographical concentrations of specific ideas.* Event-detection technology targeted at a particular area can detect either hard or soft events in a particular vicinity, such as a sensitive transport terminus, a stadium or a public event.
- *Serendipitous detections.* An important and identified cause of intelligence failure, and indeed a cognitive bias more generally, is for the collection and analysis of information to be designed around a preconceived body of expectations and beliefs. Untargeted event detection can function as a bulwark against this tendency, working to serendipitously pick out 'unknown unknowns' for further analysis.
- *An integrated situational awareness observatory.* A suite of different deployments of some of the explored event-detection methods and associated technology could interact and cooperate with each other. Trends, for instance, noticed by general untargeted systems could focus the parameters of more targeted systems for further analysis. This more complex and sophisticated deployment could take the form of an instituted general observatory, attempting to reduce the surprise of a range of different kinds of intelligence shocks.

Ethics of event-detection technology

The collection, analysis and use of any kind of social media information are ethically contested and controversial issues. It is important that any deployment of event-detection technology occurs within a framework that can balance the possible harm that they may entail against the good that they may confer.

The notion of 'harm', with regard to the collection and use of data, rests on a key distinction: whether the data relate to an identifiable individual, and whether that individual could have had a reasonable expectation of privacy regarding that information.[31] If data are 'personal' – that is, personally identifying and/or private – their collection, retention and storage must be conducted according to one of a number of legitimising legal, professional, regulatory and ethical frameworks and codes.[32]

The two event-detection methods here avoid many of these ethical and legal difficulties. The data it collects and uses are not personal, and they are placed into the public sphere by the choice of Twitter users. They are anonymised, aggregated and, in any case, subject to a very low reasonable expectation of privacy. It is generally agreed that Twitter data are in the public domain. Twitter's terms of service and privacy policy state: 'What you say on Twitter may be viewed all around the world instantly'[33] and 'We encourage and permit broad re-use of Content. The Twitter API exists to enable this.'[34]

However, this does not mean that event-detection technology entails no legal or moral hazard. The technology carries two primary moral hazards. First, the sustained collection and collation of even publicly available information about an individual or group can result in obtaining private information through, for example, the construction of a profile or 'pattern of life'. For public agencies, this constitutes 'directed covert surveillance' and must be authorised and overseen under RIPA. The extent to which the use of data by event-detection technology falls into this category depends on whether, through the targeting of the system, it is likely to build a profile of a group of identifiable individuals over time. If it does, it must be carried in compliance with the measures set out by RIPA.

Second, there is also the wider societal question of the extent to which the constant collection of even anonymous and aggregated data, especially for the purposes of security, should become routine. Event detection as a form of 'soft surveillance' can nonetheless have a chilling effect on the free expression and association integral to social media, with potential negative social and economic repercussions for the country as a whole.[35]

To guard against this, we argue that any deployment of event-detection technology must be animated by the following principles in order to retain legitimacy[36]:

- whatever the framework of its use, to be carried out on a publicly argued footing;
- to have sufficient, sustainable cause: that every use is for one of a body of recognised legitimate reasons, and that its use only continues for as long as that reason is sustainable;
- to be explicit and public in stating the research aims and methods used where possible;
- to consider whether the measures taken might reasonably be viewed as proportionate and necessary by those whose data are potentially being monitored, and whether the measures could be defended as such;
- to assess if any of the measures might undermine the existence of a free and open Internet, and in turn would cause damage to the economic and social wellbeing of the nation;
- to assess whether such measures are an effective use of public money.

Conclusion

Finding ways to maintain situational awareness is an ever-present, ever-changing challenge.[37] It has always been and remains a keystone capability of society's effective response to terrorism, but, as the world changes, the methods and sources that are needed to do this also change. As one CIA agent remarked to another, transfixed by the fall of the Berlin Wall live on television, at 'a time when public political activity proceeds at such a rapid and fulminating pace, the work of agents is overtaken by events publicly recorded'.[38] The world we live in is liable to systemic shocks and unforeseen but consequential events, but it is also a society that records more about itself, and reports more about events, than ever before. In this context it is important for event-detection technology that leverages these new, copious veins of information, like Twitter – from Olympic Medals and urban riots – now to be considered for the purpose of countering terrorism. While nascent and experimental, the methods presented in this chapter constitute a step towards intelligence work fulfilling its enduring mission in a new context: situational awareness in the age of social media.

Notes

1. Described in Stuart Allen, *Citizen Witnessing* (Cambridge: Polity Press, 2013), pp.1–4.
2. Twitter's role in breaking the story of the killing of Osama bin Laden goes beyond the role played by Sohaib Athar. The news of Bin Laden's death broke on Twitter before it was announced by President Obama. See Matt Rosoff, 'Twitter Just Had its CNN Moment', *Business Insider*, 2 May 2011, http://www.businessinsider.com/twitter-just-had-its-cnn-moment-2011-5.
3. Her Majesty's Inspectorate of the Constabulary, *The Rules of Engagement: A Review of the August 2011 Disorders* (London: Crown Copyright, 2011), http://www.hmic.gov.uk/media/a-review-of-the-august-2011-disorders-20111220.pdf, p.36.
4. 'Epidemic Intelligence: Systematic Event Detection', World Health Organization website, http://www.who.int/csr/alertresponse/epidemicintelligence/en/.
5. James R. Shea, 'Winnowing Wheat from Chaff: Tracking Down Soviet Underground Nuclear Explosions', *CIA Historical Document Release* (1996) Vol. 13, No. 4, https://www.cia.gov/library/center-for-the-study-of-intelligence/kent-csi/vol13no4/html/v13i4a03p_0001.htm.
6. Thereby avoiding a number of measurement biases often present during direct solicitation of social information, including memory bias, questioner bias and social acceptability bias. Social media is, by contrast, often a completely unmediated spectacle.
7. Described in Allen, *Citizen Witnessing*, p.9.
8. For an analysis of the usefulness of the twitcident following the 22 May killing of Lee Rigby in Woolwich, London, see Jamie Bartlett and Carl Miller, *@metpoliceuk* (London: Demos, 2013).
9. Often described as the 'wheat and chaff problem', see Mark M. Lowenthal, *Intelligence: From Secrets to Policy, Fifth Edition* (Washington, DC: CQ Press College, 2011), p.84. See also the key paradigm used in Nate Silver, *The Signal and the Noise: The Science and Art of Prediction* (London: Penguin, 2012).
10. Simon Fodden, 'Anatomy of a Tweet: Metadata on Twitter', *Slaw*, 17 November 2011, http://www.slaw.ca/2011/11/17/the-anatomy-of-a-tweet-metadata-on-twitter/.
11. 'event, n.', *OED Online*, June 2013, http://www.oed.com/view/Entry/65287?isAdvanced=false&result=1&rskey=skCBUi&.

12. Arkaitz Zubiaga et al., 'Towards Real-Time Summarization of Scheduled Events from Twitter Streams', *Proceedings of the 23rd ACM Conference on Hypertext and Social Media*, http://nlp.uned.es/~damiano/pdf/zubiaga2012summarization.pdf.
13. The interval essentially defines the scale of the event to which the algorithm is sensitive; smaller intervals allow for brief events and larger intervals allow for longer-lasting events to be detected. For example, if an interval size of 5 minutes was chosen, the rate of tweets in the last 5 minutes (t_0) is compared with the rate from the 5 minutes before that (t_{-1}). This can be done at a frequency that is independent of the interval itself, so the two sequential 5-minute segments (amounting to the last 10 minutes of tweets) could be checked every second. This allows for a more timely reporting capability than the granularity of the rate increase interval.
14. $t_0/t_{-1} = d$. So if there were 200 documents matching the filter in t_{-1}, and 500 in t_0, the differential would be $(500/200 =) 2.5$.
15. The rate increase threshold defines how rapidly the response to the event must increase for it to be detected. Lower values allow for the response to be gradual and higher values require that the reaction is more timely.
16. The chance (or probability) of a term is estimated from a sample of tweets by dividing the number of tweets that the term occurs in by the total number of tweets in the sample.
17. The difference in the two probabilities is ascertained using an information theoretic measure called Kullback–Leibler Divergence (KLD), which measures the difference in these two probability distributions as a measure of relative entropy. More precisely, a 'point-wise' version of KLD is used because the full-term probability distributions are not accessible. Pointwise $KLD(t) = P_b(t) * \log(P_b(t)/P_l(t))$, where $P_b(t)$ and $P_l(t)$ are the probabilities of the term in the background and local distributions, respectively.
18. 'Medallists – 200m Butterfly Men', Olympic website, http://www.olympic.org/swimming-200m-butterfly-men; the event was scheduled between 19:30 and 21:20. See 'London 2012 Olympics: Swimming Schedule', *The Telegraph*, 15 February 2011, http://www.telegraph.co.uk/sport/olympics/8325908/London-2012-Olympics-Swimming-schedule.html.
19. Her Majesty's Inspectorate of the Constabulary, *The Rules of Engagement: A Review of the August 2011 Disorders*, especially, pp.36–39.
20. See David Omand, Jamie Bartlett and Carl Miller, *#Intelligence* (London: Demos, 2012)
21. This could also be described as a body of terms similarly behaving in an unusual way. The relationship between this behaviour and events is taken from Weng et al., 'Event Detection in Twitter', Fifth International AAAI Conference on Weblogs and Social Media, 17–21 July 2011.
22. A signal, a mathematical tool for understanding the behaviour of elements of the Twitter-stream, is a function that conveys significant information about the behaviour of the term as part of the tweet-stream.
23. This burstiness measure is DF–IDF (document frequency–inverse document frequency), a variant of a well-known information extraction metric know as TF–IDF (term frequency–Inverse document frequency). The first part of DF-IDF, the document frequency, is simply the likelihood of the term in the current interval of tweets, as seen in the KLD calculation previously. The second part, the inverse document frequency, is the logarithm of the total tweets in the historical window divided by the number of tweets that contain the term. The burstiness is greatest

when the term appears in many of the tweets in the current interval, and few in the historical interval.
24. The Discrete Wavelet Transform algorithm is a signal processing technique similar to the Fourier transform in that it captures information in the frequency domain, but it differs in that it also captures information in the temporal domain.
25. Normalisation refers to the process of scaling the absolute values from all of the scales with regard to the sum total of the values. This allows the measurements of variability at the different scales to be comparable to each other. This also has the ancillary benefit that the result holds to all of the constraints of a probability mass function, which allows for the principled measurement of entropy over those values.
26. Pair-wise similarity: the similarity of stage 2 signals is measured by calculating the cross-correlation between two signals. Cross-correlation is a signal-processing technique for measuring the similarity between two signals. Here, un-normalised cross-correlation with a zero time lag is used.
27. A modularity-based graph-partitioning algorithm is used to recognise clusters.
28. The sum of the similarities between the terms in the cluster is used as a score. These scores are weighted by the size of the cluster. Large clusters are heavily penalised due to a factorial denominator in the penalisation function.
29. Bartlett and Miller, *@metpoliceuk*.
30. The New York City Police Department, 'Radicalization in the West: The Homegrown Threat', 2007, http://www.nypdshield.org/public/SiteFiles/documents/NYPD_Report-Radicalization_in_the_West.pdf.
31. 'Data Protection Technical Guidance Determining What is Personal Data', *Information Commissioners' Office*, 21 August 2007, http://www.ico.org.uk/upload/documents/library/data_protection/detailed_specialist_guides/personal_data_flow chart_v1_with_preface001.pdf.
32. For social researchers, this would typically be the Economic and Social Research Council, *Framework for Research Ethics Updated 2012* (London: ESRC, 2012), http://www.esrc.ac.uk/_images/Framework-for-Research-Ethics_tcm8-4586.pdf. For public officials, this would typically be RIPA.
33. 'Terms of Service', *Twitter*, https://twitter.com/tos.
34. 'Privacy Policy', *Twitter*, https://twitter.com/privacy.
35. Omand, Bartlett and Miller, *#Intelligence*.
36. These principles are drawn from Omand, Bartlett and Miller, *#Intelligence*.
37. Joseph E. Roop, *Foreign Broadcast Information Service, History Part 1: 1941–1947* (Washington, DC: Center for the Study of Intelligence, 1969), http://www.foia.cia.gov/sites/default/files/FBIS_history_part1.pdf.
38. Cody Burke, 'Freeing Knowledge, Telling Secrets: Open Source Intelligence and Development', *CEWCES Research Papers* (2007), No. 11, p.18, http://www.international-relations.com/rp/FreeingKnowledge.pdf.

Jihad Online: What Militant Groups Say About Themselves and What It Means for Counterterrorism Strategy

John C. Amble

> I say that there are two sides in the struggle: one side is the global Crusader alliance with the Zionist Jews, led by America, Britain, and Israel, and the other side is the Islamic world.[1]
>
> Osama bin Laden

In 2012, 11,098 people were killed in terrorist attacks worldwide.[2] In 2011, terrorist violence left 12,533 dead.[3] Since 2008, nearly 70,000 have died as victims of terrorism, and more than twice that many have been wounded.[4] To be sure, it is not the sole threat to global security, but terrorism remains among the most intransigent of such challenges, in part because of the magnitude of these casualty figures, but equally because of the secretive nature of its perpetrators and its consequent unpredictability. While the ideological motivations underpinning acts of terrorism are diverse, jihadists pose the most visible and most geographically extensive threat. As such, and despite the emergence and evolution of other challenges, jihadist terrorism remains a top priority of government security institutions.[5]

Countering this threat of jihadist violence requires an adherence to the maxim that has guided strategic thinking throughout centuries of conflict: that knowing one's adversary is critical to achieving victory. But in their efforts to respond to the security threats posed by jihadist groups, governments have not sufficiently realised this crucial objective. Identifying the reasons for this failure goes beyond the scope of this chapter. Nevertheless, there are concrete steps that can be taken to rectify this shortcoming. This chapter examines one such step: making full use of information available in open sources to conduct a rigorous and systematic examination of jihadist groups' media productions.

The quote that introduced this chapter represents an effort on the part of Osama bin Laden to characterise the conflict between al-Qaeda and like-minded jihadists, and the world order against which they align themselves as both global and binary. And indeed it is global: attacks executed by jihadist groups have been witnessed in virtually all corners of the world. But the binary nature that bin Laden sought to convey with his statement is superficial. It ascribes a uniform quality to the global jihadist movement atop which al-Qaeda symbolically sits. Such a monolithic conceptualisation is inaccurate. Discerning the nuanced distinctions that separate the many regional jihadist groups that form this movement has proved to be an elusive task. To identify these subtle but crucial differences – to not just know an adversary but to know many adversaries – requires a holistic and integrated approach to collecting and analysing intelligence about these organisations. A variety of tools and methods should be incorporated, but a vital contribution must come from a comprehensive and systematic examination of their open source propaganda statements and media releases.

Al-Qaeda under bin Laden's leadership developed an expert sense of how to use media to advance its goals. Jihadist groups worldwide have emulated this strategy. Each day, new text-based statements and multimedia productions are injected into the global media domain. They might be delivered to journalists or traditional media outlets, published on a militant group's own website, passed through a network of external sympathisers, posted to media-sharing platforms or spread via social media outlets. While these statements can be extraordinarily effective as tools of self-promotion and with which to shape the information environment, they also provide an opportunity for counterterrorism practitioners to peel back superficial layers and find clues that facilitate the development of a deeper understanding of each group. Carried further, these clues can aid in developing a strategic means of combating the threats that these various groups pose.

A variety of factors differentiate jihadist combatant groups from one another, but chief among them is the necessity for each group to reflect local realities in order to maintain relevance within the regions where they operate. There is simply no means by which any jihadist group can avoid being shaped by the local context out of which it emerges. Like the broader political Islamism of which jihadists represent the most violent expression, jihadism is best understood as a universal ideology that manifests itself in highly unique localised forms.[6] So while the rhetoric employed by various jihadist militant organisations around the world invokes common themes, beneath the surface can be found the subtle clues that indicate both the magnitude and the nature of the threat that each group poses. Identifying such clues is supremely important to the task of countering these threats, and will rely to an important degree on comprehensive OSINT collection and analysis.

Counterterrorism and open source information

Governments have long collected information from open sources and analysed it in order to extract its intelligence value. The US Army's OSINT field manual begins by highlighting the country's long history of exploiting open sources in pursuit of its aims: 'American military professionals have collected, translated, and studied articles, books, and periodicals to gain knowledge and understanding of foreign lands and armies for over 200 years.'[7] But one of the earliest attempts to institutionalise the collection and analysis of information exclusively from open sources came in the UK with the 1939 formation of BBC Monitoring, tasked with monitoring radio broadcasts associated with the Axis powers. The US followed suit in 1941 with the creation of the Foreign Broadcast Monitoring Service (later renamed the FBIS).[8]

Both agencies continue to serve as their countries' primary means of exploiting open source intelligence (even when the CIA launched its Open Source Center in 2005, it was built around the existing FBIS structure). These organisations, however, were created and took their form in an era markedly different in two fundamental ways from the present operational environment. The first of these centres on the nature of conflict itself. World War II and the Cold War, in both of which BBC Monitoring and FBIS played valuable roles, pitted powerful states against one another and relegated non-state actors largely to the periphery. Today, however, OSINT practitioners operate within a paradigm of conflict in which non-state actors form a central part of the picture, a significant fundamental shift from the conditions that shaped their formation. Of course, this change impacts all intelligence practitioners, not just those charged with an OSINT mandate. But the more traditional clandestine intelligence agencies have undergone significant transformations in order to align their operations with new security challenges to a degree that those focusing on OSINT have not. The creation of the Open Source Center in 2005 did not significantly alter the fundamental premises on which OSINT practitioners operate so much as expand FBIS's existent operations in recognition of the comparatively greater role of OSINT in an era defined to a considerable degree by the threat of terrorism.

The second change, however, affects OSINT services particularly acutely. The media landscape that characterised the period in which BBC Monitoring and FBIS were formed was vastly different from that of today. The 'one-to-many' communication model of the so-called 'old media', characterised by its very limited set of communicators that convey information to a wide audience, has been replaced by a 'many-to-many' model, whereby innovative new tools give every information receiver the capability to also act as an information transmitter. As a result, the number of sources available to OSINT practitioners has grown exponentially. This, combined with the transformation of the shape of conflict, means that the task required of OSINT practitioners – to support efforts to counter the threat of terrorism – is one for which their organic, operational evolution has not fully prepared them.

One of the most effective ways in which OSINT can contribute to a comprehensive counterterrorism strategy is through the systematic collection and analysis of jihadist groups' media. Such an effort's independent value should not be overstated, but neither should those of more technical or more covert intelligence disciplines. Individual pieces of intelligence, regardless of the nature of their source, must be collected and synthesised within an all-source construct in order to develop the most complete picture of security threats possible. The notion of a single piece of information providing actionable intelligence absent the context provided by a range of sources – the 'ticking time bomb' scenario – might make for compelling drama in film and spy novels but it does not reflect reality. For a number of reasons, not least of which is the sheer quantity of information available, OSINT offers many of these critically important contextual pieces of intelligence. In the conflict with jihadist groups, those clues that emerge from an examination of the groups' media can be vital.

Collecting and analysing jihadist media, however, is not without challenges. Most obvious among these is a function of the same quality that gives OSINT its value – the magnitude of the universe of open sources and the amount of information that they produce. Even focusing exclusively on statements released by militant groups' official media wings requires the monitoring of a much larger set of intermediate sources, due to the variety of dissemination methods that jihadists employ. The task of sifting through tens of thousands of statements, audio recordings and videos in order to determine the meaning and value of each is a supremely difficult one. But this is not a challenge unique to OSINT. Intelligence services face it to an enormously greater degree when collecting and analysing SIGINT. The same enterprise data-management tools that help SIGINT-focused agencies to overcome this obstacle can be applied to the problem of collecting, analysing and organising jihadist groups' external communications.

A greater challenge is posed by the prevalence of disinformation and misinformation, which similarly exists across a range of intelligence disciplines but represents an especially difficult problem in OSINT generally and when examining jihadist media in particular. Jihadist groups are well aware that their adversaries will monitor their statements, and their communications are designed in part with this audience in mind. Analysts must thus seek to isolate individual kernels of accurate information from the mountains of what is conveyed as truth but ranges from mildly distorted to wholly inaccurate. The problem of misinformation is compounded by the anonymity of the virtual domain in which most jihadist media are now disseminated, where sources purporting to speak on behalf of a particular group might have no such connection, and further so by the dynamics that characterise the global community of online jihadists, which encourage such posturing as a means of boosting one's credibility among one's peers.

The difficulties of assessing the veracity of information, however, can also be mitigated considerably by adopting existent methods from other intelligence disciplines. The accuracy of information contained in jihadist media can be graded according to a variety of factors, similar to the way in which agencies treat individual HUMINT reports. Specific sources can also be coded to reflect reliability, which can be continually assessed based on source placement and access, and historical accuracy of information.

The obstacles to leveraging jihadist media as an intelligence tool are real and, if not addressed in a systematic way, they can deeply erode the usefulness of such an effort. But there are practical steps that can be taken to mitigate these challenges. Beyond these concrete steps, the value of these open sources of intelligence can be most effectively harnessed by developing and bringing to bear deep bases of theological, cultural, social and political understanding. By doing so, and by effectively leveraging jihadist media as an OSINT tool with which to appreciate threats to regional stability and global security, governments can enhance their capacity to evaluate each of the many jihadist groups that are active around the world, particularly with respect to two principal characteristics: strategic goals, and operational and tactical planning.

Jihadist groups' strategic objectives

In late May 2013, coordinated suicide bombings struck two targets in the West African country of Niger, killing more than 20 people.[9] One of the attacks targeted an army barracks in the city of Agadez. The second was aimed at a uranium mine near Arlit, 240 kilometres to the north. Immediately after the deadly bombings, a statement by a militant commander named Mokhtar Belmokhtar was sent to a Mauritanian news agency and posted to jihadist web forums in which he claimed that the attacks were jointly conducted by his al-Mua'qi'oon Biddam ('Those Who Sign in Blood') group and the Movement for Oneness and Jihad in West Africa (known by its French acronym, MUJAO).[10] Significantly, both of these groups are led by and largely comprise fighters who have strong ties to al-Qaeda in the Islamic Maghreb (AQIM).

AQIM fighters and those of the group's numerous splinter factions and affiliates have also been active in neighbouring Mali. One of the most prominent groups fighting in the north of that country in the wake of a 2012 coup that destabilised the country is Ansar al-Din, a large contingent of whose fighters are either current or former members of AQIM. MUJAO has also actively fought in Mali, mainly around the city of Gao. As with the Niger bombings, Belmokhtar's al-Mua'qi'oon Biddam has worked closely with MUJAO in this area.[11] Each of these three groups was formed after 2011, and each retains ties to the core leadership of AQIM. In fact, the groups are best understood as partners that operate within the loose network that continues to spread across a growing expanse of North and West Africa but that retains AQIM as its central hub.

At first glance, both AQIM's interest in Mali and Niger and the organisational transformations that have seen AQIM-linked figures instrumental in leading militant groups that are active across a growing swathe of the region seem surprising. AQIM's lineage can be traced back to the bloody civil war that engulfed Algeria throughout much of the 1990s.[12] Prior to a pledge of allegiance to Osama bin Laden in 2006 and a subsequent name change in 2007, AQIM was known as the Salafist Group for Preaching and Combat. This group itself was formed in 1998 by a commander of the Armed Islamic Group (known as the GIA, its French acronym), which was the leading opposition group fighting against the government during the civil war. Thus AQIM's evolutionary origins as a unified combatant organisation are firmly rooted in a highly nationalistic Algerian context. However, these nationalist roots are apparently becoming increasingly peripheral.

AQIM has historically made use of traditional trade routes across the Sahel for logistical functions and to engage in smuggling activities that finance its operations. But since 2011, the wider region has gone from one of seemingly secondary importance to one in which the group has been increasingly operationally active. In addition to the operations in Niger and Mali perpetrated by groups with strong links to the al-Qaeda affiliate, the organisation has also forged ties with Nigerian jihadist group Boko Haram,[13] and reports also indicate that Belmokhtar was possibly involved in the planning of a September 2012 attack against the US consulate in Benghazi, Libya, that left four Americans dead, including the US ambassador to the country.[14] Taken together, these events form a picture of AQIM's increasingly pan-regional set of strategic objectives.

These developments, while surprising given AQIM's Algerian nationalist roots, were portended in the group's media releases. A systematic approach to analysing these media releases, available via a number of open sources, could have warned of the newly emergent regionalised threat. Some of the evidence of their increasingly expansive strategy is clear, such as that provided by a February 2011 statement addressed to the people of Libya, in which AQIM encouraged Libyans to 'continue their jihad and revolution', and called on Muslims around the world to support the fighters who were battling the forces of Gadhafi.[15] A statement released a month earlier and addressed to the people of Tunisia expressed a similar message.[16]

A commitment to directing OSINT analytical capabilities towards systematically assessing AQIM's media during this period could also have identified more subtle signs of its increasingly pan-regional strategic outlook. In December 2011, al-Andalus Foundation, the group's media wing, released a 35-minute video entitled 'A Speech to our Amazigh Brothers'.[17] That the recorded speech was delivered in Amazigh, a language spoken by ethnic Berbers across a vast range of territory from North Africa's Mediterranean coast through the southern Sahel, is significant for two reasons. First, AQIM media has traditionally been disseminated almost exclusively in Arabic, with a minority also being released in French. Thus the break in this longstanding pattern represented by the Amazigh video suggests that expanding the group's influence among this minority ethnic community has become a particular strategic objective. But, more importantly, the relationship between Algerian Arabs, who form the core of AQIM's membership, and the country's Berber minority is a tense one. These deep-seated tensions have a long history. During the war for Algerian independence from France, the French exploited them by raising irregular units from among the Berber population to fight against the predominantly Arab anti-colonial National Liberation Front.[18] Given this historical context, AQIM's efforts to reach out to Amazigh-speaking Berbers hints at a change in the group's strategic calculus. Whether the group independently decided to expand its area of operations more aggressively into regions populated by Amazigh speakers or simply sought to exploit a spontaneously apparent opportunity when places such as northern Mali underwent significant upheaval is largely irrelevant. In either case, the outcome is the same: AQIM and its satellite factions were to become more active across a larger region. And, significantly, this pattern was presaged in the group's media productions that were freely available to intelligence analysts in the open source domain.

The internal group dynamics that led to AQIM's transformation from a unified combatant organisation to a central node within a network of splinter groups is also important. The case of Belmokhtar is instructive. Formerly one of AQIM's most prominent brigade commanders, his relationship with the core leadership cadre surrounding the group's emir, Abu Musab Abdel Wadoud, has historically been troubled.[19] The differences that led him to establish al-Mua'qi'oon Biddam are manifold, but in a statement announcing his new group's formation, he promised to focus on the kidnapping of foreigners, a particular specialty of the AQIM brigade that he headed.[20] This does not prove that differences over the targeting of foreigners were a primary factor in the creation of his semi-autonomous group. But, given his statements, his involvement in recent attacks in a number of countries, and his status as one of the few senior North African jihadists who fought in Afghanistan in the early 1990s, his outlook does appear to be significantly more universal than those of some other AQIM leaders, and it is certainly more so than that which was predominant in AQIM's formative years.

Identifying the extent to which jihadist groups and their leadership cadres are primarily locally, regionally or globally oriented is a difficult task. Regardless of orientation, jihadist groups typically emulate the same universal themes put forward by al-Qaeda's core, at least to some extent. This is true of AQIM, particularly since its 2006 oath of allegiance to al-Qaeda. For instance, the North African syndicate frequently refers to its adversaries as 'crusaders', an emotionally evocative term and the label with which al-Qaeda routinely defines the West. Of 199 media productions released by AQIM between 2007 and 2011, a staggering 107 of them included some form of the Arabic word for 'crusade'.²¹

But much of this rhetoric is superficial. Among the most deeply entrenched pillars of global jihadist thought is the distinction between the near enemy (*al-adou al-qareeb*) and the far enemy (*al-adou al-baeed*), which has its roots in the writings of Egyptian Islamist Sayyid Qutb, a leading figure of the Muslim Brotherhood in the middle of the twentieth century.²² For al-Qaeda, the far enemy is represented by an alliance of Western governments, but pride of place at the centre of this alliance is reserved for the US and Israel. For AQIM, however, France has consistently been characterised as the group's principal far enemy. This is a result of the legacy of French colonialism, and is an indicator of AQIM's awareness that tapping into emotive, lingering anti-French grievances maximises the group's relevance among local constituencies. Belmokhtar, on the other hand, seems comparatively less motivated by such considerations. While he has vocally criticised French involvement in the region, he has similarly denounced that of other foreign countries. Significantly, his perception of the far enemy appears to be more universal than that of AQIM's core leadership. Identifying the degrees of difference in the outlooks of individual jihadists and the groups that they lead offers a means of developing tailored strategic responses, and OSINT analysis of AQIM's media provides an illustrative example of this point.

AQIM is not alone among jihadist groups whose leadership has experienced schisms over strategic objectives. In many cases the divisive question goes beyond that of who, exactly, constitutes the most appropriate far enemy, and centres instead on the related but much more fundamental issue of whether a group's strategic decision-making should be guided by national, regional or universal objectives – whether the local focus that defines the formative stages of many groups should be retained or set aside in favour of a more expansionist set of goals. In these cases, too, a jihadist group's media releases offer important clues as to its strategic outlook, the identification of which is central to determining the extent and the nature of the threat that it poses. Until his death, Osama bin Laden served as the public face of al-Qaeda. His imprimatur on the group's media releases lent credibility, but frequent references to him even in those releases in which he did not directly feature implicitly confirmed the organisation's continued commitment to his strategic direction. This holds true for regional groups as well. The choice of which members of an organisation's leadership cadre are most frequently highlighted is strongly indicative of which strategic viewpoints are dominant at any given time.

Between 2008 and 2011, Somali jihadist group al-Shabaab's prolific media operatives produced and disseminated an impressive total of 563 text, audio and video releases. Over this period, the relative prominence of various members of the organisation's core leadership in its media productions has ebbed and flowed. Traditionally, the group's emir, Sheikh Mukhtar Abu Zubeyr, earned the most frequent references. By 2010, however, another senior official, Sheikh Mukhtar Robow, was also increasingly highlighted, earning nearly as much publicity in al-Shabaab media as Abu Zubeyr. And in 2011, a third figure, Sheikh Hassan Dahir Aweys, became increasingly prominent after merging the Hizbul Islam group that he led with al-Shabaab.[23]

These trends are significant. Al-Shabaab is the product of a long series of splits within successive Islamist organisations in Somalia. One such group, al-Itihaad al-Islamiya, was formed during the unceasing civil war that wracked the country throughout the 1990s and quickly became Somalia's most prominent Islamist combatant group. The group dissolved in the early 2000s, and several of its key figures were later instrumental in the formation of the Islamic Courts Union (ICU), a judicial entity that restored a degree of order to parts of Mogadishu and employed armed militias to enforce the courts' decisions. These militias also facilitated the ICU's expansion into other parts of southern Somalia. By the end of 2006, however, the courts network splintered after successive defeats by Ethiopian forces stripped the ICU of most of its territory. Al-Shabaab was formally born when an ICU armed wing known as Harakat al-Shabaab al-Mujahideen ('the Mujahideen Youth Movement') subsequently emerged as an independent organisation.[24]

But al-Shabaab is not immune to the persistently divisive dynamics that seem to plague Somali Islamist groups. The group has recently seen a schism develop that similarly threatens to break it apart. At the heart of the division are two groups of al-Shabaab leaders, one that seeks to hold to its long-term emphasis on establishing an Islamic state in Somalia, and another that advocates an increasingly international focus.[25] Aweys and Robow are prominent among the former group; Abu Zubeyr leads the latter. Aweys' and Robow's growing prominence in al-Shabaab media in 2010 and 2011 was related to the group's concerted effort to build support with tribal leaders, with whom the duo's comparatively more nationalist orientation was most likely to resonate. However, examination of the group's more recent media reveals a subsequent decrease in references to the two leaders compared with those to Abu Zubeyr.[26] This suggests the probability that Abu Zubeyr's strategic objectives – more universal than those shared by Aweys and Robow – were at that time growing in influence within the organisation. Here again, OSINT analysts' systematic evaluation of jihadist groups' media offers a valuable means of identifying the leaders who are most prominently featured and thus determining which strategic outlook is on the ascendant. This determination is fundamental to forecasting the nature of the threat that each group poses to both regional and international security.

Identifying strategic shifts is the first step towards forecasting their consequences and determining whether such consequences will be felt principally within a particular country, across a wider region or truly globally. More importantly, identifying these shifts is necessary in order to formulate approaches aimed at limiting the impact of these consequences. By examining and evaluating the open source media releases of regional jihadist groups, their secretive inner workings can be penetrated, and an understanding of the strategic outlook that determines their future course can be both developed and refined.

Operational plans and tactical execution

While identifying how regional jihadist groups are strategically orientated is essential to conceptualising the long-term threat that they each pose, detecting tactical adaptations and shifts in operational planning serves a more immediate purpose by equipping military forces and intelligence agencies engaged in combating them to develop effective countermeasures. Employing OSINT examination of media releases contributes greatly to this task as well.

Tactical innovation occurs rapidly in warfare, and the global conflict between jihadist organisations and their state-based adversaries is no exception. During the US-led war in Iraq, an action/reaction pattern took shape in which both new and innovative weapons and rapidly fielded countermeasures meant to limit these weapons' effectiveness were constantly introduced into the battle space. Increasingly deadly roadside bombs such as explosively formed projectiles (EFPs) triggered the development of the mine-resistant ambush protected (MRAP) vehicle. Improvised rocket-assisted mortars, which a top US general claimed in 2008 represented the greatest threat then faced by coalition forces, led to a fundamental re-evaluation of base security procedures.[27]

For years, Iraq effectively served as a laboratory for emerging tactical innovations, essentially equivalent to a military proving ground in an active combat zone. The global attention focused on the country also made it important not just for the disparate set of groups that operated there but also for regionally oriented jihadist organisations around the world. Thousands of foreign fighters flocked to Iraq to join militant groups waging war against coalition forces. Those who survived and returned to their countries of origin brought back with them detailed technical knowledge of the latest battlefield innovations.

The proliferation of improvised explosive device (IED) technology provides a telling example of this pattern. For heavily outgunned militant groups in Iraq, IEDs served as a deadly equaliser. As a result of the effectiveness demonstrated by the homemade weapon in Iraq, it was also adopted by a rapidly growing array of groups outside the country. Between 2008 and 2011, IED attacks outside the combat zones of Iraq and Afghanistan more than doubled, with worldwide totals averaging more than 600 attacks per month by early 2011.[28] Not only has the weapon proliferated, but technical innovations that significantly increased its killing power in Iraq have spread to other countries as well. EFPs – shaped charge devices that propel molten copper projectiles at velocities sufficient to penetrate even heavily armoured vehicles – were so dangerous that they contributed to the US Defense Department's decision to procure and deploy tens of billions of dollars worth of MRAP vehicles.[29] That technology subsequently spread, and has been utilised by combatant groups in other regions.[30] Naturally, the likelihood of such a diffusion of technical knowledge is amplified by the existence of both personal and organisational relationships between a regional militant group and those in the country from where the tactical innovation is imported.

Identifying these relationships and determining whether they are sufficient to facilitate the process of importing tactical innovations is a task that OSINT analysts examining jihadist media releases are well placed to accomplish. Al-Shabaab, like other regional jihadist groups, frequently seeks to use its media releases to place its conflict in the broader context of a global war between Muslims and their oppressors. Thus references to Iraq, Afghanistan, Chechnya, Palestine and other conflict zones are common. But al-Shabaab appears to hold particular reverence for jihadists in Iraq. In 2008 the group named its newly formed media organisation the al-Zarqawi Centre for Studies and Research in honour of Abu Musab al-Zarqawi, the slain former leader of al-Qaeda's Iraq affiliate, the Islamic State of Iraq (ISI).[31] In May 2010 the group issued an audio message eulogising two other ISI leaders who had recently been killed.[32] A week later, al-Shabaab released a video publicising a suicide car bomb attack on a Mogadishu base housing soldiers from the African Union Mission in Somalia (AMISOM), declaring the action to be revenge for the two ISI leaders' deaths.[33] Al-Shabaab's evident virtual devotion to the ISI – which is not replicated to the same degree with respect to other groups – suggests an increased probability that al-Shabaab would be among the earliest adopters of tactics developed and refined in Iraq. Indeed, Somalia was one of the first countries to see the employment of the EFP technology that was used heavily against Iraqi security forces and coalition troops deployed in the country.

Suicide bombings represent another tactic increasingly utilised by jihadist groups around the world. The frequency of such actions has risen dramatically over the past two decades. Between 1994 and 2000, there were an average of 22 suicide attacks worldwide annually. By 2004, that number had jumped to 119. Between 2005 and 2011, the average yearly total leapt to 221, a tenfold increase over the figure from just over a decade earlier.[34]

Suicide attacks provide yet another means for jihadist groups to mitigate their typical comparative disadvantage vis-à-vis their adversaries, in terms of resources and weaponry. They are attractive principally for the simple reason that they are highly effective – serving essentially as inexpensive and reliable guided munitions. Terrorists also recognise the psychological impact of suicide bombings on their adversaries. But the corollary to this psychological effect is an aversion to the tactic's use among communities that jihadist groups rely on for support. This is particularly important when regional jihadist groups are understood as occupants of a blurred zone between traditional terrorist organisations and insurgencies.[35] As terrorists, suicide bombings magnify the terror that these groups seek to sow; as insurgents, they risk diminishing the popular support critical to their existence. The theological debate surrounding the morality of suicide attacks with respect to Islam is well documented.[36] However, some groups face greater cultural opposition as well. In such cases, regional jihadist groups face the imperative of overcoming this opposition if they wish to employ the effective tactic.

To accomplish this task, many jihadist groups have turned to their media operatives. Jihadist media embraces the concept of martyrdom to link the actions of suicide bombers to an emotionally and historically evocative sense of dying for one's faith and in defence of one's coreligionists. Suicide bombers are celebrated as heroes. Thus references to martyrs and martyrdom are frequent in media releases that are produced by groups that choose to increasingly rely on suicide attacks and are intended to persuade local constituencies of the acceptability of the tactic. Pakistan provides a telling example of this pattern. Until the last decade, suicide attacks in the country were exceptionally rare. Only two such attacks were reported in 2002, and only six as late as 2006. From there, the number jumped dramatically, with 55 reported in 2007 and dozens more each year since.[37] According to a study by the Pew Research Center, however, support for suicide bombings among Pakistanis was shown to be very low, with only 5 per cent expressing approval for the tactic in 2008, just as they were becoming increasingly common in the country.[38]

In an effort to diminish opposition to the tactic, groups active in Pakistan have used their media releases to lionise suicide bombers, conveying the notion that they represent the heroic defenders of a Muslim population under siege. The Islamic Movement of Uzbekistan (IMU) provides an example. The jihadist group originated in Central Asia, subsequently moved to Afghanistan, and relocated to Pakistan after the US-led invasion that toppled the Taliban regime. Since its move to Pakistan, it has increasingly aimed its attacks at targets inside the country. Analysis of releases from the group's organic media wing, Jundullah Studios, shows that 60 per cent of its communiqués issued between 2009 and 2011 contained words that translate to 'martyr' or 'martyrdom' (the IMU produces media in multiple languages).[39] Thus, OSINT analysts examining a regional jihadist group's media are among the best situated to discern signs of emerging tactical patterns, such as a heightened emphasis on suicide attacks, and to assess the future likelihood of such attacks.

Operationally, the way in which a jihadist organisation employs media can also assist in identifying shifts in emphasis on various areas of operation within the country or region in which the group is active. Regional jihadist groups, even those that incorporate nationalist strands into their guiding ideologies and operate almost exclusively within a single country, rarely enjoy equal freedom of movement throughout the country. This is the case even in the security vacuum that exists in the most failed states. In Somalia, for years the epitome of state failure, al-Shabaab has been unable to exert its control uniformly throughout the country. Instead it has been confined to specific areas by its adversaries: primarily by Ethiopian, Kenyan, AMISOM forces, and to a lesser extent by Somali government troops, and in some cases by tribal militias. But the group's operational geography is dynamic, with the areas in which it is most active shifting constantly.

After its splinter from the ICU, al-Shabaab quickly took over control of much of Mogadishu, subsequently spreading to other parts of southern Somalia. Since then, however, areas under its control and in which it operates have been fluid. The group has regularly shifted its emphasis between various administrative regions in the country. Signs from al-Shabaab media have presaged these shifts. In early August 2011, al-Shabaab abruptly withdrew all of its forces from Mogadishu.[40] The fighters moved into other regions in southern Somalia under the group's control, but also into more contested territory in the country's central region in which it had not previously maintained a strong presence – namely, the Hiraan and Galguduud administrative regions. In the six months preceding the move, al-Shabaab disseminated eight media releases that highlighted its interest and activities in the two regions – an annualised average of 16 media productions devoted to Hiraan and Galguduud. In the 40 months prior to that period, the group produced ten such releases – an annualised average of just three.[41] The numbers involved in these calculations are too small to serve as an independently reliable metric, but the figures do provide an indicator of al-Shabaab's heightened interest in central Somalia before their move there, one that could have been detected by OSINT practitioners by taking a comprehensive macroanalytical approach to examining jihadist media.

Similarly, analysis of al-Shabaab's media releases can yield more immediate signs of impending threats to security. Rather than moving into a new area by force, the group often seeks to persuade or coerce local tribal leaders or militias into aligning themselves with al-Shabaab. It then leverages this expanded network to grow its presence in the new area, after which it can conduct attacks itself or induce its new allies to do so on its behalf. In late February 2012, al-Shabaab issued a communiqué announcing that the leader of a local militia operating in a mountainous area of Puntland, a semi-autonomous region in northeastern Somalia, had sworn allegiance to al-Shabaab's emir on behalf of his fighters.[42] Less than a week later, two further releases announced that the militia had executed their first attacks as al-Shabaab's anointed affiliate in Puntland, an area that had previously been largely free from the organisation's violence.[43]

Hence, just as regional jihadist groups' media releases can provide indicators of strategic geographic orientation across a broader region, as was seen with AQIM's expansion into neighbouring countries, it can also assist in forecasting operational shifts into new regions within a particular country. As those charged with the mandate of collecting and analysing information from the sources in the public domain, OSINT practitioners thus play a critical role in developing and refining a picture of jihadist groups' operational footprint that provides the necessary context within which the respective threats that they pose must be assessed.

Conclusion

Countering the collective threat that jihadist groups pose to regional stability and global security has emerged as arguably the most challenging contemporary task confronting security agencies. Two features make it particularly difficult. First, jihadist groups have demonstrated resilience and a collective capability to evolve rapidly. The dangers associated with this process of evolution change equally rapidly, both in nature and in magnitude.

Second, while the threat posed by jihadist organisations is in fact global, they do not form a homogenous body readily countered by a single strategic approach. Each group emerges out of a distinct set of conditions and is influenced by highly localised dynamics. For each of them, these dynamics shape its identity, help to determine its strategic aspirations, and impose limits on the extent to which it can truly adhere to universal themes of global jihadism.

Incorporating rigorous examination of jihadist groups' media releases into standard OSINT practices, however, can help to overcome the difficulties associated with these two factors. To be sure, the independent utility of such analysis should not be exaggerated. At the risk of stating the obvious, no single panacea exists to eliminate the threats posed by jihadist groups operating around the world. But open sources such as these are readily accessible, they can provide considerable value, and they are in many ways absent complications and risks that can arise with other categories of intelligence sources. Alongside the intended messages in jihadist groups' media also sit clues that facilitate the development of a critical degree of understanding about militants' strategic objectives, operational outlook, and the tactical approaches that they adopt to sow violence and threaten stability and security. Ultimately, crafting flexible, strategic measures to counter these threats requires a holistic approach that leverages a comprehensive set of tools. But given the nature of terrorism and the requirements of countering it, OSINT offers opportunities to help to craft strategic responses to the threats posed by jihadist groups that have yet to be fully exploited. A robust,

Notes

1. Bruce Bennett Lawrence, *Messages to the World: The Statements of Osama bin Laden*, translated by James Howarth (London: Verso, 2005), p.73.
2. 'Country Reports on Terrorism 2012: Annex of Statistical Information', *United States Department of State*, p.3, http://www.state.gov/documents/organization/210288.pdf.
3. 'Country Reports on Terrorism 2011: Annex of Statistical Information', *United States Department of State*, p.2, http://www.state.gov/documents/organization/195768.pdf.
4. Ibid.
5. For instance, US Director of National Intelligence James R. Clapper highlighted terrorism as one of the global threats confronting the US's intelligence community as recently as March 2013. James Clapper, 'Statement for the Record: Worldwide Threat Assessment of the US Intelligence Community', Senate Select Committee on Intelligence, 12 March 2013, pp.3–5, http://www.intelligence.senate.gov/130312/clapper.pdf.
6. Anders Strindberg and Mats Wärn, *Islamism* (Cambridge and Malden, MA: Polity Press, 2011), p.9.
7. US Army, *FMI 2–22.9: Open Source Intelligence* (Washington, DC: Headquarters, Department of the Army, 2006), p.1.
8. For an examination of early US OSINT efforts, see J. E. Roop, *Foreign Broadcast Information Service, History Part 1: 1941–1947* (Washington, DC: Center for the Study of Intelligence, 1969).
9. Adam Nossiter, 'Suicide Bombings in Niger Kill Dozens in Dual Strikes', *The New York Times*, 23 May 2013, http://www.nytimes.com/2013/05/24/world/africa/niger-hit-by-two-suicide-attacks.html?_r=0.
10. 'Mokhtar Belmokhtar "Masterminded" Niger Suicide Bombs', BBC, 24 May 2013, http://www.bbc.co.uk/news/world-africa-22654584.
11. 'AQIM and Its Allies in Mali', *The Washington Institute for Near East Policy*, 5 February 2013, http://www.washingtoninstitute.org/policy-analysis/view/aqim-and-its-allies-in-mali.
12. For a detailed examination of AQIM's lineage, see James D. Le Sueur, *Between Terror and Democracy: Algeria Since 1989* (London and New York: Zed Books, 2010), pp.122–168.
13. Jacob Zenn, 'Boko Haram's International Connections', *CTC Sentinel* (2013), Vol. 6, No. 1, pp.7–13.
14. 'Phone Call Links Benghazi Attack to al Qaeda Commander', CNN, 5 March 2013, http://www.cnn.com/2013/03/05/world/africa/benghazi-al-qaeda.
15. AQIM media release, 'Support and Backing for the [Libyan] Revolution of our Family, the Free, Descendants of "Umar al-Mukhtār"', 23 February 2011.
16. AQIM media release, 'To Our People in Tunisia', 26 January 2011.
17. AQIM media release, 'A Speech to Our Amazigh Brothers', 9 December 2011.
18. Alastair Horne, *A Savage War of Peace: Algeria 1954–1962*, Fourth Edition (New York: New York Review of Books, 2006), p.256.

19. 'Entretien Exclusif avec Khaled Abou Al-Abass, alias "Balouar" ', *Agence Nouakchott d'Information*, http://www.ani.mr/?menuLink=9bf31c7ff062936a96d3c8bd1f8f2ff3&idNews=15829.
20. 'Belmokhtar Breaks Away from AQIM', *Magharebia*, 11 December 2012, http://magharebia.com/en_GB/articles/awi/features/2012/12/11/feature-03.
21. The quantitative analysis of media releases is based on a comprehensive collection that I have compiled from multiple sources.
22. Seth G. Jones, *Hunting in the Shadows: The Pursuit of Al Qa'ida Since 9/11* (New York and London: W. W. Norton & Company, 2012), pp.43–44.
23. Robow featured in twice as many media releases in both 2010 and 2011 than in 2009. Aweys was never mentioned until the December 2010 merger of his group with al-Shabaab but featured in 11 releases over the subsequent year.
24. For further reading on the history of jihadist groups in Somalia, see Gregory Alonso Pirio, *The African Jihad: Bin Laden's Quest for the Horn of Africa* (Trenton, NJ and Asmara, Eritrea: The Red Sea Press, 2007).
25. Muhyadin Ahmed Roble, 'Al-Shabaab Split Threatens to Open New Conflict Between Somalia's Islamists', *Jamestown Foundation Terrorism Monitor* (2012), Vol. 10, No. 9, pp.5–6.
26. In the first three months of 2012, as the rift between the two leadership factions grew, neither Robow nor Aweys featured in any al-Shabaab media releases.
27. 'U.S. Forces Pursue Iraq Insurgents' Top Bomb Secrets', *USA Today*, 11 July 2008. IRAMs, also known as lob bombs, are rocket-propelled metal canisters, often empty propane tanks, filled with explosives and shrapnel, and typically fired from the backs of trucks.
28. Peter W. Singer, 'The Evolution of the IED', *Armed Forces Journal*, February 2012, http://www.brookings.edu/research/articles/2012/02/improvised-explosive-devices-singer.
29. Alex Rogers, 'The MRAP: Brilliant Buy, or Billions Wasted?', *TIME*, 2 October 2012, http://nation.time.com/2012/10/02/the-mrap-brilliant-buy-or-billions-wasted/.
30. 'Remarks by Lieutenant General Michael D. Barbero', Director, Joint IED Defeat Organization, 14 August 2012, https://www.jieddo.mil/content/docs/20120814_LTG_Barbero_Remarks_ONR_asPrepared.pdf.
31. Al-Shabaab media release, 'Young Mujahideen Movement Media Department Presents the First Issue of the Magazine, "The Faith of Abraham" ', 3 October 2008.
32. Al-Shabaab media release, 'The Islamic State Will Remain', 2 May 2010.
33. Al-Shabaab media release, 'The Islamic State Will Remain, with Permission from Allah', 8 May 2010.
34. Data from the National Consortium for the Study of Terrorism and Responses to Terrorism (START) Global Terrorism Database (GTD), retrieved from http://www.start.umd.edu/gtd.
35. This conceptualisation of terrorist groups is best articulated by David Kilcullen, *The Accidental Guerrilla: Fighting Small Wars in the Midst of a Big One* (London: Hurst & Company, 2009).
36. For a discussion of this, see, for instance, Scott Atran, 'The Moral Logic and Growth of Suicide Terrorism', *Washington Quarterly* (2006), Vol. 29, No. 2, pp.127–147; and Robert A. Pape, *Dying to Win: The Strategic Logic of Suicide Terrorism* (New York: Random House, 2005).
37. Data from the National Consortium for START GTD.

38. 'Unfavorable Views of Jews and Muslims on the Increase in Europe', *Pew Research Center*, 17 September 2008, http://www.pewglobal.org/files/2008/09/Pew-2008-Pew-Global-Attitudes-Report-3-September-17-2pm.pdf.
39. The quantitative analysis of media releases is based on a comprehensive collection that I have compiled from multiple sources.
40. 'Al Shabaab's Withdrawal from Mogadishu', *American Enterprise Institute*, 7 August 2011, http://www.criticalthreats.org/somalia/zimmerman-shabaab-retreat-mogadishu-august-7-2011.
41. The quantitative analysis of media releases is based on a comprehensive collection that I have compiled from multiple sources.
42. Al-Shabaab media release, 'The Mujahideen in the Golis Mountains in Eastern Somalia Joined the Shabaab al-Mujahideen Movement', 27 February 2012.
43. Al-Shabaab media release, 'An Attack Against an Apostate Military Base Near the City of Bosaso', 3 March 2012; Al-Shabaab media release, 'The Killing of More than 30 Apostates Near the City of Bosaso', 3 March 2012.

Conclusion

Christopher Hobbs, Matthew Moran and Daniel Salisbury

In the Introduction we used the example of the recent chemical weapons attacks in Syria to illustrate the uses and value of OSINT. We found the reliance of major world powers on open sources – video, social media and journalistic reports, among others – to support their intelligence assessments in Syria striking, if not particularly surprising. For as the various chapters making up this volume have shown, the expansion of the Internet and the information revolution that this has provoked, combined with the emergence of new approaches and methodologies, mean that there are now relatively few areas that open source research cannot penetrate.

As the advance of technology continues unabated, society has embraced the Internet and the notion of connectivity; more and more of our social interactions and functions are conducted online. Moreover, this process is circular and self-reinforcing: 'smart technology' that aims to better connect us to the objects of our interests and make relevant information more accessible to us is adopted as fast as it is developed, further immersing us in the online world of the Internet.

Much of the information generated by this shift is publicly accessible, and this expansive information landscape offers a wealth of opportunities. From policing to the study of proliferation, and from responding to humanitarian crises or other large-scale emergencies to counterterrorism, processes of technological and analytical innovation have turned OSINT into a force multiplier in the all-source intelligence cycle. Not only can OSINT provide important contextual information to analysts but it is now increasingly a source of real-time, operational information. SOCMINT, for example, the new sphere of analysis proposed by Omand, Miller and Bartlett (Chapter 2), provides us with a highly innovative analytical approach to understanding social phenomena as they develop and evolve, as well as the motivations that underpin collective action in particular contexts. Miller and Wibberley (Chapter 8) also demonstrate how social media can play a critical role at the edge of practice: social media analytics can serve as a stable source of situational awareness in rapidly changing and fluid situations.

While this book has, to some extent, charted the changing role of OSINT in an institutional sense – OSINT has come to occupy a more central position in state-level intelligence efforts, for example – this is not its primary objective. Rather, the focus has been to provide readers with an insight into the dynamic research environment that has emerged under the broader theme of OSINT. The contributors to this volume have clearly demonstrated the increasing value of OSINT across a variety of fields and disciplines. At the same time, however, the various authors stress that the evolution of OSINT is a formative process; the full potential of OSINT has not yet been tapped.

Of course, while OSINT offers many benefits and advantages, it also poses significant challenges. Paterson and Chappell (Chapter 3), for example, discuss some of the security issues that our embrace of technology and the Internet has engendered. With more and more of our information online, we have become easier targets for cybercriminals who would do us harm. Open source information can also suffer in terms of credibility, particularly with online information, where there exist numerous examples of inaccuracies and disinformation.[1] Omand, Miller and Bartlett (Chapter 2) discuss this in the context of social media research, an area where strategies for validating the credibility of findings are yet to be commonly applied. Such difficulties are also seen when considering more specific topic areas. For example, Amble (Chapter 9) has considered the disinformation efforts of some jihadist groups, and Bruls and Dorn (Chapter 7) have written about the difficulties in this regard posed by humanitarian actors with particular interests. However, in many senses, similar challenges are found in relation to verifying information obtained through more covert intelligence efforts.

Another major challenge lies in the ethical dimensions of OSINT. The enormous controversy that followed revelations regarding PRISM – the secret surveillance programme put in place by the US to gather massive amounts of personal data and electronic communications – highlights a growing concern regarding the loss of privacy in a technology-driven world. OSINT certainly does not raise issues of transparency and ethics to the same extent as other, covert forms of intelligence. Indeed, a major benefit of OSINT lies in its ability to facilitate interagency and intergovernmental information-sharing and to maintain the neutrality of international organisations. As Hobbs and Moran (Chapter 4) mention, Tehran has accused the IAEA of applying 'intelligence driven safeguards', in reference to its use of third-party information in assessing Iran's nuclear activities. Here the greater use of open source information in forming its conclusions can assist in maintaining the IAEA's political credibility.[2] Similarly, we have seen how the use of open source can facilitate a variety of actors – both civil and military – in their responses to humanitarian crises. However, emerging OSINT methodologies can also bring their own ethical issues.

The authors dealing with social media analysis (chapters 2 and 8), in particular, highlight the risks associated with gathering for analysis the thoughts and emotions expressed by private individuals on fora such as Facebook and Twitter. Indeed, these attempt to grapple with these issues, presenting a number of criteria which the authors feel that open source collection and analysis efforts should meet – it is argued that such programmes must be implemented on a publically argued footing, must be in aid of a sustainable cause and must be proportionate. Clearly, however, the current rate of technological advancement means that this is a fluid and rapidly evolving area, and because legislation is struggling to catch up, such ethical maxims must be given considerably more thought.

This need for critical thinking touches on another theme, implicit in all of the contributions and explicitly mentioned in some: the importance of further exploiting the new avenues of research and inquiry that recent developments in OSINT have opened up. This book has allowed contributors to explore a range of opportunities, challenges and approaches posed by technological and methodological developments relevant to OSINT from different fields and perspectives. However, there is much more to be done. The chapters offered here should therefore be considered as a basis for further research, rather than the presentation of any conclusive and final results. It is hoped by the editors that this book will thus contribute to stimulating further work in this exciting and emergent area of research and practice.

Notes

1. Pallab Ghosh, 'Warning Sounded on the Web's Future', *BBC News*, 15 September 2008, http://news.bbc.co.uk/1/hi/7613201.stm.
2. 'Iran Prepares for Moscow', *Asia Times*, 9 June 2012, http://www.atimes.com/atimes/Middle_East/NF09Ak01.html.

Index

Additional Protocol (AP), xii, 71, 78, 79
Afghanistan, 125, 128, 129, 136, 137, 143, 144, 174, 177, 179
   NATO operations in, 126, 134
   UN Assistance Mission in, xiv, 126
all-source intelligence, 1–2, 16, 24, 35, 66, 73, 84, 136, 171, 185
al-Qaeda, 5, 168–9, 172–5, 178
   in the Islamic Maghreb, 172–5, 181, 182, 183
   links with Boko Haram, 173
al-Shabaab, 178, 180–1
Amble, John, C., vi, ix, 5, 168, 186
Anonymous (Hacktivist), 53, 62
Armed Islamic Group (GIA), xii, 173
Aspin-Brown Commission, 10, 22

Bartlett, Jamie, ix, 4, 23, 24, 40, 165, 166, 167, 185
BBC Monitoring, 14, 17, 170
Bharatiya Janata Party (BJP), xii, 76
big data, 16–17, 29–33, 35, 149
Bin Laden, 10, 147, 168–9, 173, 175
biological weapons, 81, 83, 94
Boko Haram, 5, 173
British Broadcasting Company (BBC), xii, 14, 16, 170
   monitoring, 14, 17, 170
Bruls, Fred, ix, 5, 141, 142, 143, 144, 186

California Coastal Records Project (CCRP), xii, 56
Central Intelligence Agency (CIA), xii, 10, 15, 164, 170
Centre for Science and Security Studies (CSSS), viii
Chappell, James v, ix, 4, 44, 186
chatter, 11, 150, 154, 156
chemical weapons, 2–4, 81, 83, 94, 185
   Syrian chemical weapons 4, 6, 185
China, 45, 66, 68, 89, 107
civil-military cooperation, 137
Cyber Intelligence Sharing and Protection Act (CISPA), 58, 62

closed sources, 11, 15–16, 66, 87
cloud services, 44–6, 48, 53–4, 56, 60
Cold War, 10, 14, 21, 99, 170
Communications Intelligence (COMINT), 134
   and humanitarian crises, 134
Community Open Source Program, 11
Comprehensive Safeguards Agreement (CSA), *see* safeguards
connectivity, 1, 19, 46, 136, 185
counter-terrorism, 5, 10, 145, 147–8, 168–9, 171, 185
covert intelligence, 3, 84, 89, 96, 171, 186
crime, 21, 25, 28, 32, 34–6, 47, 128–9, 131
   cybercrime, 47
   organised crime, 47
crowd-sourcing, 109, 114, 112–14
cyber attack, 48–55, 57
cybercrime, *see* crime
cybercriminal, 44, 46, 47, 52, 53, 55, 57, 59, 60, 61, 186
cybersecurity, v, xi, 4, 44, 57–8

Digital Shadows, 48, 53, 56–60, 148
Director of Central Intelligence (DCI), xii, 12
Director of National Intelligence (DNI), xii, 2
   office of the, 20, 22
Disarmament, 67
   United Nations Institute for Disarmament Research, 120
Distributed denial of Service (DDOS), xii, 47
domain name system, 49
Dorn, A. Walter, v, ix, 5, 123, 141, 143, 186
doxing, 53
Dropbox, 45, 46, 60
dual-use, 65, 67, 75
Dulles, Allen, 9–11

education, 51, 110, 111
Electronic Intelligence (ELINT), 134
and humanitarian crises, 134
enrichment, *see* uranium
European Union, 58, 81, 89
EUROPOL, 10, 22

Facebook, 18–20, 24, 28–9, 32, 34, 44–5, 56, 103, 113, 116, 135, 148
Flickr, 29, 113, 135
Foreign Broadcast Information Service (FBIS), xii, 14, 23, 167, 170

Gadhafi, Moammar, 76, 173
GIA, *see* Armed Islamic Group
Gibson, Stevyn D., v, x, 1, 4, 22, 23
government, 2–3, 10–11, 19, 47, 58, 66, 69–70, 72, 75–6, 82, 84–9, 91–2, 94–6, 107–8, 110–11, 113–14, 124, 135–40, 168–9, 172–3, 175, 180
Global Terrorism Database (GTD), xii, 183

hackers, *see* hacking
hacking, 47, 49–50
Hobbs, Christopher, v, vi, x, 1, 65, 185, 186
Human Intelligence (HUMINT), 16, 20, 27, 35, 134
and counter-terrorism, 172
issues, 134
Humanitarian crisis, 4–5, 103–18, 123–5, 128, 130–1, 135, 139, 185–6
Human Security Intelligence (HSI), 5, 7, 123, 127, 133, 134, 136–40

illicit procurement, 69, 81–2, 84, 88, 91
illicit trade, 81–4, 88, 90–4, 96
 see also illicit procurement
Imagery Intelligence (IMINT), 134
improvised explosive device (IED), 177
Information and Communication Technology (ICT), 12, 15, 19–20, 44, 135
and big data, 17–19
Intelligence assessments, 2–3, 185
Intelligence community, 1–2, 4, 10–12, 20–1, 27, 38, 65–6, 130, 135–6, 138–9

Intelligence cycle, 2, 17, 95, 185
Intelligence requirement, 14
priority, 12
Intelligence failure, 65–6, 84, 163
International Atomic Energy Agency (IAEA), xiii, 5, 66, 70–7, 78, 79, 80, 93–4, 186
Board of Governors, 71, 79, 80, 93
Department of Safeguards, *see* safeguards
International Committee of the Red Cross (ICRC), xiii, 141
Internet Protocol (IP), xiii, 50, 137
Islamic Courts Union (ICU), xiii, 176
Islamic Movement of Uzbekistan (IMU), xiii, 179
Iran, x, 5, 71, 75, 79, 80, 81, 87, 89
illicit procurement, 85, 87, 91
Iranian Revolution, 77
nuclear activities, 75, 81
transfer of technology to, 82
UN Panel of Experts on, 92–4, 97
UNSCR on Iran, 92
Iraq, 3, 66, 71, 84, 177–8
Islamic State of, 178

jihad, 168, 173–4
jihadist, 5–6, 36, 168–9, 171–82, 186
Joint Intelligence Committee (JIC), xiii, 3, 6, 36
Joint Mission Analysis Centre (JMAC), xiii, 136
Joint Operations Centre (JOC), xiii, 136
Joint Terrorism Analysis Centre (JTAC), xiii, 36

Khan, A.Q., 65, 74, 76, 84, 97, 99, 100
Kennan, George, F., 10, 22
Kent, Randolph, v, x, 5, 103

Libya, 74, 75, 76, 173, 182
London 2012, *see* Olympics

Measurements and Signature Intelligence (MASINT), 134
Medecins Sans Frontieres (MSF), xiii, 116

media, 1–2, 5, 14–15, 19, 66, 70, 75, 116, 118, 129, 138, 147–8, 150
  jihadist, 168–82
  social, 1–2, 4–5, 12, 17, 19–20, 24–38, 44–5, 48, 51–5, 60, 69, 113, 115, 117–18, 135, 147–9, 151, 155, 163–5, 185–6
  state, 87
Miller, Carl, v, vi, ix, x, 4, 5, 23, 24, 40, 147, 166, 167, 185, 186
Ministry of Defence, 14, 22, 55, 62
Missile Technology Control Regime (MTCR), xiii, 85
MI6, 56, 62, 81, 97
mobile communication, 1, 110, 114, 116
Moran, Matthew, v, vi, x, 1, 65, 185, 186

National Police Coordination Centre (NPoCC), xiii, 36
North Atlantic Treaty Organisation (NATO), xiii, 9, 22, 100
  intelligence analysts, 131
  operations in Afghanistan, 126, 129, 134
  peace support operations, 140
NMAP, 49
noise (intelligence), 5, 33, 103–4, 116–17, 149, 159
non-government organisation, 2, 66, 82, 105–6, 108–9, 111, 114, 125, 127, 135, 137–8, 140
North Korea, 5, 20, 81, 82, 92–3, 98, 99
  UN Panel of Experts on, 92–4, 98
NPT, *see* Treaty on the Non-proliferation of Nuclear Weapons
nuclear materials, 72, 93
Nuclear programme, 72, 74, 76, 81
Nuclear proliferation, 4–5, 65–7, 73, 84
nuclear weapons, 5, 15, 65–71, 75–6, 147, 162–3
Nuclear Suppliers Group (NSG), xiii, 85, 95, 96, 100

Olympics, 24, 150, 153, 154, 166
  Olympic medals, 150, 154, 162, 165
Omand, Sir David, v, ix, x, 4, 24, 38, 40, 43, 166, 167, 185, 186
online piracy, 58
Organisation for the Prohibition of Chemical Weapons, xiii, 2

Open Source Center, xiii, 2, 12, 14
open source intelligence (OSINT), 1–5, 9–11, 13–17, 20, 44, 49, 63, 65, 81–2, 103, 108, 134, 170, 185–7
  and counter-proliferation, 65–6, 69–70, 72–7, 81
  and counter-terrorism, 147, 169–82
  and cybersecurity, 44, 48, 53
  and humanitarian crises, 103–4, 106–7, 112–16, 118, 123, 134–5, 137, 139–40
  issues, 4–5, 11–12, 21

Pakistan, 74, 147, 179
Paterson, Alastair, v, xi, 4, 44, 186
peacekeeping operations, 125–7, 130, 140
plutonium, 68
  reprocessing, 67
police, 13, 14, 24, 25, 28, 36, 40, 42, 127, 155, 157, 160, 162
  Cleveland police, 14
  Metropolitan police, 24, 28, 40, 155, 162, 165
  New York Police Department, 162, 167
  Police Act (1997), 37
  UK Police constabularies, 25, 36
Police Act (1997), *see* police
PRISM, 40, 186
privacy, 14, 18–19, 37–8, 52, 59, 116, 163, 186
  invasion of, 56
  loss of, 186
  privacy controls, 27
  right to, 37, 38
private sector, 5, 10, 12, 24, 26, 69, 82, 90–2, 95, 108, 111, 113
  and illicit trade, 84–8, 90, 96, *see also* illicit trade
  organisations, 10, 25, 54
  SOCMINT and the, 26
private security company, 125, 138
protest, 24, 45

radicalisation, 43, 162
RAND, 12, 22, 40
Regional Information Collection Centre (RICC), xiv, 74
Regulation of Investigatory Powers Act 2000, xi, 37, 38, 164, 167

reprocessing, *see* plutonium
riot, vii, 28, 162, 165
  food riots, 133
  London riots, 155, 157
  Twitter and, 155, 157, 163, 175
  urban riots, 165
RIPA, *see* Regulation of Investigatory Powers Act 2000

safeguards, v, 65, 66, 70–8
  Comprehensive Safeguards Agreement, 70, 71
  IAEA Department of, 71, 73, 74, 75, 77, 79
Salisbury, Daniel, v, vi, xi, 1, 5, 98, 185
sanctions, 68, 81–2, 85, 87, 89, 92–4
Sawers, Sir John, 56, 97
Signals Intelligence (SIGINT), 14, 134, 171
  and humanitarian crises, 134
  situational awareness, 16, 29, 25, 27, 123, 126, 147–8, 162–4
Social Media Intelligence (SOCMINT), xiv, 36, 38–9, 185
  and big data, 29, 31–3
  capabilities, 24, 27, 31, 34
  concept of, 4, 20, 24–7, 185
  and counter-terrorism, 27–8
  early detection, 29
  issues with, 33–5, 37
  legitimate use of, 36, 37, 383, 39
Somalia, 108–9, 176, 178, 180
South Korea, 75
spear-phishing, 52
Stop Online Piracy Act (SOPA), xiv, 58
Structured Query Language (SQL), xiv, 150
Syria, 2, 4, 6, 103, 185
  Syrian chemical weapons, *see* chemical weapons

Technical intelligence (TECHINT), 14, 20
terrorism, 4–5, 25, 147, 162, 164, 168, 170, 181
  online terrorism, 4, 28
  state-sponsored, 76, 124
  terrorist, 5–6, 28, 89, 115, 129, 147, 162, 168, 179

terrorist, *see* terrorism
Treaty on the Non-proliferation of Nuclear Weapons, xiii, 65, 70, 71, 78, 79, 80
Tweet, *see* Twitter
Twitter, 18, 19, 23, 41, 45, 116, 150, 157, 167, 187
  analytical potential of, 28, 30, 135, 163, 165
  event detection with, vi, xi, 29, 147, 149–62, 166
  feeds, 31, 113
  rise of, 1, 6, 44, 60, 103, 148
  Tweet, 1, 28
  Twitter and electoral campaigns, 32, 42
  use in Haiti, 116

United Kingdom, 2–3, 24–5, 28, 67, 74, 86, 91, 115, 118, 128, 139
  authorities, 13
  intelligence community, 12–14
  Ministry of Defence, *see* Ministry of Defence
United Nations, 70, 81–2, 85, 89, 92–4, 105–6, 108, 109, 116–17, 124–7, 131, 133–4, 136
  Assistance Mission in Afghanistan, *see* Afghanistan
  development program, 131, 133
  security council, 92–3
United States, 14–15, 27, 45, 116, 125, 138, 147, 173, 175, 186
  Bureau of Industry and Security, 98
  Commission on Secrecy, 10, 22
  intelligence community, 1–3, 10, 12, 22, 40, 65, 182
  legislation, 58, 62
uranium, 68, 81, 87, 172
  enrichment of, 67, 75
Ushahidi, 103, 113

Weapons of Mass Destruction (WMD), 67, 76, 81, 83, 85, 87, 89, 96

Youtube, 19, 29, 97, 113

ZMAP, 49, 61

CPSIA information can be obtained at www.ICGtesting.com
Printed in the USA
LVOW01*1122070715

445253LV00010B/374/P

9 781137 353313